FOUR
THOUSAND
HOOKS

FOUR
THOUSAND
HOOKS

A True Story of Fishing
and Coming of Age
on the High Seas of Alaska

© 2012 by the University of Washington Press

16 15 14 13   5 4 3 2

All rights reserved. No part of this publication may
be reproduced or transmitted in any form or by any means,
electronic or mechanical, including photocopy, recording,
or any information storage or retrieval system,
without permission in writing from the publisher.

University of Washington Press
PO Box 50096, Seattle, WA 98145, USA
www.washington.edu/uwpress

Printed and bound in the United States of America
Designed by Ashley Saleeba
Composed in Warnock Pro; display type set in Cubano

Please visit fourthousandhooks.com

Library of Congress Cataloging-in-Publication Data
Adams, Dean J.
Four thousand hooks : a true story of fishing and coming of age
on the high seas of Alaska / Dean J. Adams.
pages cm
ISBN 978-0-295-99197-9 (cloth : alk. paper)
1. Adams, Dean J. 2. Fishers—United States—Biography.
3. Fisheries—Alaska. 4. Alaska—Description and travel.
I. Title. II. Title: 4,000 hooks.
SH20.A36A3 2012   639.2092—dc23   2012019070

Diagrams drawn by Joan Forsberg
Map drawn by Pat Grant

The paper used in this publication is acid-free and
meets the minimum requirements of American National
Standard for Information Sciences—Permanence of Paper
for Printed Library Materials, ANSI Z39.48-1984.∞

FOR TWO WOMEN AND FIVE MEN

DIANE AND LORI
JACK, FREDDY, KAARE, WALLY, AND CHRIS

CONTENTS

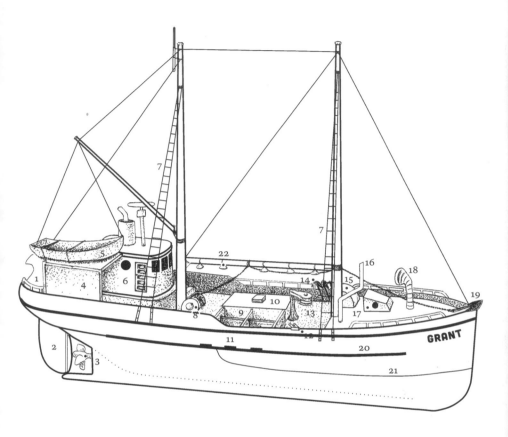

Profile diagram of the *Grant*

1	Chute	13	Gurdy
2	Rudder	14	Gear anchors on rail
3	Propeller	15	Companionway
4	Skiff support	16	Galley stove
5	Skiff		exhaust pipe
6	Captain's stateroom	17	Sky light
	(behind pilothouse)	18	Ventilation funnel
7	Rigging	19	Ship's anchor
8	Anchor winch	20	Guard
9	Checkers	21	Water line
10	Hatch cover	22	Main boom &
11	Scuppers (3)		deck lights
12	Roller		

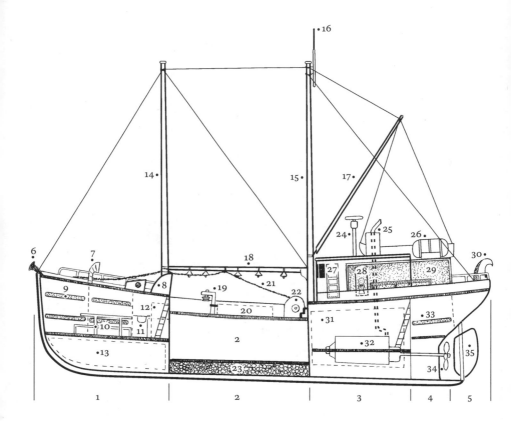

Cutaway diagram of the *Grant*

1	Fo'c'sle	13	Water tank	25	Smokestack
2	Fishhold	14	Forward mast	26	Life raft
3	Engine room	15	Aft mast	27	Pilothouse door
4	Aft cabin	16	Antenna	28	"Shitter"
5	Lazarette	17	Aft boom	29	Bait table & tent
6	Ship's anchor	18	Main boom		(shaded area)
7	Ventilation funnel	19	Gurdy & coiler seat	30	Chute
8	Companionway		(black bar)	31	Fuel tanks (4)—
	& ladder	20	Hatch combing & cover		flanking engine
9	Bunks (6)	21	Anchor chain	32	Main engine
10	Table & bench	22	Anchor winch	33	Bunks (2)
11	Sink	23	Ballast	34	Propeller
12	Stove & fridge	24	Radar	35	Rudder

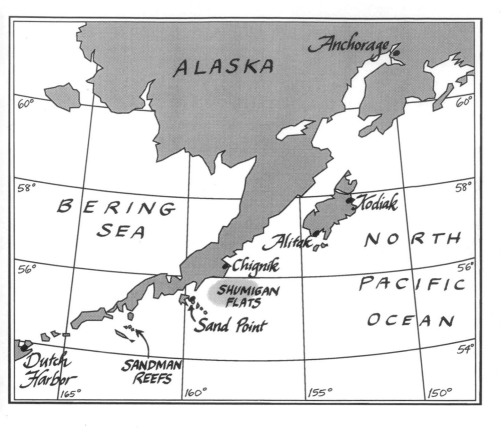

Western Gulf of Alaska and the Bering Sea

PART ONE

The Sea Is Fluid, Elusive, Hard to Grasp

1

THE *GRANT*

THE IMPACT JOLTED THE OLD SCHOONER. TIMBERS SHIVERED down the length of the vessel. In that moment, I understood the elasticity of one hundred tons of boat and cargo.

The collision had been sharp. Uncle Jack had told me, "If we hit something . . . like a log . . . take the boat out of gear. That'll keep the propeller safe." My book flipped out of my hands to the floor when I lunged for the engine control. The *Grant* slowed, cutting through the sea by momentum alone.

Just outside the railing, a huge log passed by—thirty feet long, stripped of bark, and spiked with hairy splinters. Waves sloshed over its back. Most of the log's mass was below the water's surface, like an iceberg.

Freddy's head popped up within the shelter of the fo'c'sle companionway. The ship's cook looked riled. His eyes locked onto the log, then shot up to me inside the pilothouse. I gave him a nervous giggle that he couldn't hear, shrugged my shoulders, and smiled back with a look that said, "Sorry about that."

Freddy's eyes glared at me as he descended back into the fo'c'sle.

To survey the ocean ahead, I squinted, straining to see through the salt-crusted windows.

Seven weeks had passed since I had begun my education on the Alaskan ocean, learning that logs collect in "lines of drift" at "eddies," drifting together with other "flotsam" and "jetsam"—logs, seaweed, and debris—at the borders of ocean currents.

Seven weeks ago I was fifteen years old.

Satisfied that we had clear sailing ahead, I reached up to the engine control and put the boat into forward gear, pushing the throttle handle forward. The belly of the boat was full, with 70,000 pounds of halibut packed in ice. The *Grant* labored to regain cruising speed.

On deck below the pilothouse, our 3,000-pound deckload lay hidden under layers of tarps. These days the halibut schooners rarely returned to port with full loads. The boat was heavy and lazy. It was a great feeling.

I wedged my butt onto the seat and had just bent over to pick up my book when the door burst open.

Freddy flew into the pilothouse. Shouting, he fought with the knob to Jack's stateroom door, trying to reach my sleeping uncle, "Jack! Jack! . . . hey, Jack! We're taking on water in the fo'c'sle!"

The door gave way with a loud crack. I heard my uncle spring out of his bunk and cry out, "Huh, wh . . . what!"

In his underwear, Jack dashed out the door, chasing after Freddy down into the fo'c'sle. In less than a minute he was back up on deck, racing to the pilothouse. His face was drawn so tight his lips were white. He reached past me, groping for the engine control and took the propeller out of gear. Then he wheeled around, stopped with a jerk, reached down to the floor to grab my book, and hurled it over the side of the boat into the ocean.

Without a word to me, he ducked into his stateroom and snatched up his clothes. Feet shot through pant legs and arms through sleeves. He slipped on his boots and dashed out the door, this time running to the stern.

Chris and Wally appeared from the fo'c'sle, roused from sleeping in their bunks.

With the boat out of gear, I guessed that I was relieved of duty and skulked up forward to see what was happening in the fo'c'sle.

I stuck my head down the companionway and looked below at the same time as a rush of water spilled over the galley's floorboards. Water was pouring from the bilge into the fo'c'sle at a volume so great that even I knew our pumps couldn't handle it. Instinctively, I muttered "Oh, shit!"

the traditional oath of the wayward voyager. Chris suddenly was beside me, peering down. My mind raced.

"Can't we just find the hole and plug it?" I asked him.

"Nope. The hull's double-planked. The hole's in the outer planking and we can't find it down there, 'cause the inner planking's in the way. There's no way to get at it."

Behind me, Jack appeared, returning from the engine room below. The *Grant*'s most powerful pump—the engine-driven pump—now pumped a thick, clear stream of water over the side.

Jack fired out orders: "Wally, get on the deck pump." Kaare appeared from the stern. "Kaare, here . . . get this other one going. Get it primed and start pumping . . . Chris. . . . Where's Chris?" I looked around. I was just talking to him. He seemed to have vanished.

Looking down the companionway, I could see that the flooding hadn't slowed at all. The water in the fo'c'sle was getting deeper and the boat was settling lower into the ocean. On deck, seawater was now gushing in through the scuppers to submerge the main deck under two feet of water. The cumulative effort of three pumps—the engine-driven pump and the two manual pumps—was not enough. We were losing ground.

Pumping occupied all of the crew—everyone but me. Left to watch, I faced the fact that the *Grant* was sinking. My limbs grew numb. I was scared, but I fought the feeling. I had to keep my mind clear. I had to work out a plan for saving the boat.

The flood of water obviously came from outside the boat. I reckoned that if we couldn't plug the hole from the inside of the hull, I would try to plug it from the outside. I had to locate the hole—but I'd have to go overboard to find it.

I knew that I could execute my plan, but I needed a big plug. My sleeping bag could do this, but it was in my bunk—deep in the fo'c'sle.

Peering down through the companionway, I saw that the seawater now covered the food lockers. Food and debris sloshed across the breadth of the fo'c'sle—potatoes, onions, cans and jars, clothing and bedding all writhed together in chaos below.

Thousands of gallons of seawater filled the main deck, rising up to approach the companionway's threshold. If this threshold was breached while I was down there, I knew that I'd never get out, trapped by the ensuing waterfall. Over the past weeks I'd come to fear the power of water. I had to do something and it had to be quick.

Looking down, I saw Chris's hand grabbing for the ladder. I moved to let him pass as he raced up the rungs with his wool coat clenched in his teeth. I rushed down the ladder into the fo'c'sle and jumped to the galley bench, trying not to look at the madness swirling below my feet. I reached into my bunk and grabbed my sleeping bag. On impulse, I snatched my ditty bag, containing my camera and letters from home, and I scrambled back up the ladder.

I set my ditty bag aside and turned to the forward deck, grabbing a length of heavy rope—a slipshot line, out of the flagpole rack. I dashed up onto the bow, which was still dry and above the waves. Looking down to the main deck, I saw Kaare and Wally straining at the pump handles, standing thigh-deep in water and looking weary. Hopelessness lined their faces.

Chris stood by, wearing his wool coat now, waiting to spell the others. I waved him toward me. Seeing that I was up to something, he came up to the bow. With the bow soaring and falling with the waves, we struggled to stand. I hurried to tie a knot, improvising a climbing harness by looping a thick rope around my hips.

"Keep an eye on me, Chris. I'm going over the side." From behind his glasses, I saw his eyes widen, but he didn't try to stop me.

I straddled the bow rail with one leg dangling over the side of the boat and wrapped the rope around the pipe rail to make a belay line.

My mind became focused, so clear that it calmed me. I felt good now. I was confident that I could really do this—I could save the ship.

I held both lines of the rope attached to the harness and eased my grip to let the rope slip through my hands, lowering myself down. The bow dove into a wave, sending churning water rushing up at me. I gripped the ropes tight and jerked to a halt. The boat recoiled from the wave, rising high to fall fast and hard. I looked up to Chris who stood by, clutching my sleeping bag. I wondered if I was doing the right thing.

The bucking of the boat settled down a bit and I loosened my grip on the rope, descending the white hull until giant black letters rose up to stare me in the face . . .

G R A N T . . . my grandfather's boat.

2

THE RAFT

TENSION GAVE WAY. I RELEASED MY GRIP ON THE ROPE AND descended down the bow. The *Grant* rebounded from waves, weaker now as the enormous burden of water flooding the deck slowed the boat's movement. She began to founder—half sunk, half floating.

I had hoped to have a good view of the red and white border of the boat's waterline, but it was submerged deep underwater, barely visible as I hung from the rope. I knew that the log had hit somewhere on the port side of the bow. A twenty-foot stretch of the *Grant*'s flared bow was vulnerable to forward impact. Swinging from my harness, I could cover only half that distance. Luck would have to be on my side for me to find the hole, reach it from the rope, and plug it with my sleeping bag.

I tried to inspect the hull. The boat dropped some, dipping my feet and legs into the water. Not stopping, I lowered myself down further. A large wave passed under the boat, lifting the bow higher, so that the bow sank down deep into the next wave, dunking me into the sea all the way up to my armpits. The icy water felt like fire. Breath exploded from my lungs.

My priorities changed. I scanned the bow for one final inspection, received another dunking, and retreated without a thought. Between waves, I became somewhat weightless and could climb in increments up the boat's side, boosting myself up the rope each time the boat fell. When I reached the top, Chris helped me roll over the pipe rail and pulled me back up onto the deck.

In the short time I'd been trying to save the boat, a tumult of foam and breaking waves had flooded the entire main deck. The ocean looked calm by comparison.

The deck was empty of people. Everyone had disappeared. I couldn't believe my eyes. I became lost in a fog of jumbled thoughts.

"Dean . . . Chris . . . Get back here!" Jack appeared, waving urgently from the stern.

His yelling snapped me out of my daze.

"NOW, GODDAMMIT!"

I snatched up my ditty bag.

The *Grant's* anchor chain hung below the boom that spanned the main deck. To avoid being washed overboard, I grabbed the chain hand over hand and swung to the hatch combing—an island in the chaos—then jumped to join Jack on the poop deck. Chris scrambled close behind.

The stern was tilted high in the air, levered by the sinking bow. The *Grant* seemed poised for a plunge to the seafloor, hundreds of feet below.

Approaching the stern rail, I saw the orange canopy of the life raft floating in the water below. It was tied to the *Grant* by a slender cord and looked like a pet on a leash. Bouncing in the sea, the raft looked frisky next to the sinking ship, which was rolling slower and slower now, as if weakened by the exertions of staying afloat.

Upon seeing the raft, my hope of saving the *Grant* was crushed. Jack must have given the order to abandon ship. I realized that Freddy, Wally, and Kaare had already boarded the life raft. Only Jack, Chris, and I remained on board.

Freddy's head stuck out through an opening in the raft's canopy. I saw Wally's and Kaare's legs inside the raft behind him. Freddy waved his hand, beckoning us to come.

Jack shoved a life jacket out to me, giving an order: "Put it on." I could see in his eyes that he too had lost hope of saving the *Grant*.

Reality was moving faster than my mind's ability to keep up. I lapsed into a state of shock and heard my uncle's voice as though a great distance separated us. The commands he shouted became muddled as I struggled to put on my life jacket, preparing to abandon ship—an act of lifesaving that I had never practiced, contemplated out loud, or discussed with anyone. It didn't matter what happened now. I felt the cold indifference of shame. I didn't care—because I couldn't care.

Chris went first. Freddy backed away from the raft's entryway and Chris jumped for it. He missed the raft and plunged into the ocean. He flailed wildly in the frigid water, looking as though he was attempting to fly out of the sea. Freddy's strong arms reached out and snatched Chris from the water, yanking him straight into the raft.

It was my turn.

Holding my ditty bag, I leaped from the stern and landed, sprawling over the soft, inflated circular tube that formed the perimeter of the raft. My lower body dangled into the cold ocean. I let my bag fall into the raft and groped for a handhold. I felt Freddy's fingers latch onto the scruff of my shirt, and he pulled me out of the sea like a gaffed fish. With a quick jerk of his arm, he deposited me in a heap inside the raft.

The orange canopy arched overhead, lit up bright by the sun, casting orange over the grim faces of our little group. The color saturated everything. The absence of noise struck me—gone were the pulse of the engine and the cacophony of water. No crash of waves. No deck for the water to roar across. Responding to the whim of the sea, the raft bobbed like a cork.

I don't remember Jack entering the raft and cutting the tether line with a knife. The little raft began drifting in the ocean, free from the *Grant*.

I'll never forget that glow of orange.

3

FAMILY

GRAMPA JACOB HAD ENORMOUS, GNARLED HANDS. WHEN I was a boy, my mother's father was the oldest man in the world, and it bothered me that his hands shook some—even when he wasn't doing anything. But with those same hands, he had built a tiny replica of a boat, carving it from a single block of balsa wood. Not meant as a toy, it was about a foot long, a representation of the 68-foot Pacific halibut schooner that he captained for decades of his life at sea. A tiny anchor projected from the bow, and a propeller from under the stern—each made of wood, whittled with precision. To fashion the boat's rudder, he had scavenged a tin can from the trash and cut a square from it. He bent the square of metal around a short length of wire snipped from a coat hanger, and then soldered them together. He drilled two holes into the soft wood and inserted slender dowels for the boat's masts and booms, held erect by cables of sewing thread, gossamer strands of rigging.

It was a beautiful little craft, constructed with such detail that seemed impossible considering the tremors that shook his hands. The only signs of his frail condition were the wobbly, tiny letters painted on the bow and the stern, distinguishing the schooner as his boat—the *Grant*. I reckoned that when he built it, he yearned to be back at sea, in command of his vessel. I was told he had patterned the model from memory, without any sketches, drawings, or plans whatsoever. All the dimensions of the *Grant*, three of space and one of time, were etched into his mind.

A few years later, Grampa lay prone and mute, crippled by strokes. Before he went into the nursing home, my family—my mom, my brother, and I—paid him a Sunday visit. I was about eight years old. Seeing the model, I asked him outright, "The model . . . can I have it, Grampa?" He gave me a look, reached out, and handed it over to me. My mother rolled her eyes, knowing that the model was too fragile for child's play. Grampa hadn't built it with children in mind. Looking back, I suppose now that my grandfather gave away the model, because he sensed my passion and not my want.

My little brother and I played with this heirloom for many years. The model deteriorated in our care, falling apart, dismasted, her rigging in shambles. We had no business possessing it. Sometime after my grandfather passed away, my mother stuffed the broken remnants of the model into a shoebox and put it away on a dark shelf.

Between 1910 and 1930, shipyards in the Pacific Northwest built more than 150 halibut schooners identical in design to my grandfather's boat, constructed with planks, ribs, and keels hewn from giant evergreen trees harvested from virgin forests—boats built expressly for catching halibut, the largest flounder fish in the world. Two masts and a long and narrow hull characterized the halibut schooners, as did a small pilothouse positioned on the aft of the vessel. Following its launch and maiden voyage in 1926, the *Grant* departed Seattle each year along with the other schooners in the fleet, traveling north to Alaska.

My grandfather was born in 1889, the eleventh child of Norwegian farmers. With no hope of working or inheriting the family farm, he took a job as a seaman on board a British square-rigger leaving Norway before the Great War (WWI) and sailing as far as Australia. Grampa initially settled in British Columbia, starting his fishing career delivering halibut to the port of Prince Rupert. He eventually moved to Seattle, where he spent the rest of his life.

I remember how his voice deepened to a growl when he talked about Norway—he called it "the Rock."

Grampa Jacob had three sons. Donny was the eldest, followed by Jack, then Peter. Each had learned the trade from Grampa at an early age, working on the deck of the *Grant*, leaving home for the first time after turning fourteen or fifteen. Donny was the first to go. He said he was "shanghaied" into fishing—that is, forced against his will.

When Grampa retired in 1959, he offered to sell the *Grant* to his sons at fair market value—take it or leave it—no special deals for family. At twenty-four and twenty years of age, respectively, Donny and Jack accepted his offer. Sixteen-year-old Peter was left out.

By the time I turned fifteen, Donny had sold his interest in the *Grant* and owned the schooner *Attu*. Peter had bought the schooner *Northern*. Jack stayed with the *Grant*.

All three were successful fishermen. Each year, they produced some of the largest catches in the fleet. They had earned the status of "highliners"—a hard-earned title reserved for the best fishermen and captains in the North Pacific.

My mother, Diane, was the eldest child in my grandfather's family—and the only girl. It was never discussed or considered by the family that she go north to fish. In order to do well for herself, she knew that she needed to marry well. As a young coed at the University of Washington, she met my father and they fell in love. After they married, she withdrew from the university, pregnant with me. My brother Jon followed me by twenty months.

To all appearances, my mother *had* married well—my father was tall, dark, and handsome. But in no time, my father proved himself to be an entrepreneur with a roving eye, a dreamer of ill-conceived enterprises in business as well as in extramarital love. Before I had reached kindergarten, my parents' marriage and their finances were both in shambles. They divorced, and from then on, my mother provided for her two boys on her own.

After the divorce we moved from apartment to apartment until the summer of 1964, when, through the help of a friend, my mother found a small house to rent near Lake Meridian, in the foothills of the Cascade Mountains.

A couple of years later, while I was still in elementary school, we were forced to move out of this house. It was to be knocked down to make way for a housing development. To stay in the area, we would have to downsize.

We squeezed into our new home. It was a tiny dwelling dwarfed under a stand of towering fir trees, a short walk from the shore of Lake Meridian, known to us simply as "the Lake." The walls of our small house lacked insulation. Mildew grew in its musty corners. Even so, we referred to our home with affection, calling it "the Cabin."

Our kitchen was a nook, having room for a stove and sink but not a refrigerator. The latter sat outside the front door, its motor humming away on our porch. No matter the weather—hot or cold—it was always cooler on the porch than inside the cabin. As a child, I rationalized that *all* refrigerators should be out of doors; it took less energy to keep things cool that way. Nevertheless, during the infrequent Pacific Northwest cold snaps, we suffered from seasonal inconveniences: the contents of the refrigerator—milk, eggs, and lettuce—froze solid.

4

JULY 5, 1972

WAVES OF TRAVELERS SURGED THROUGH THE CORRIDORS OF
Seattle-Tacoma International Airport. I stood in line with Mom and my
brother, shuffling my feet as I surveyed the bustling scene, waiting to
check my baggage to Alaska. I had just quit my job at an ice cream shop
to go fishing on the *Grant* with Uncle Jack. My canvas sea bag lay squat
on the floor of the terminal, overstuffed like a big sausage with all my
gear and clothing.

To get to the airport on time, we had awakened at dawn. I felt dazed.
My mother stood, erect and stoic. My fourteen-year-old brother, Jon—
younger and taller than I—stared wide-eyed, looking more nervous than
I felt. In less than an hour, our family trio would shrink to two, and I
guessed that he was preparing himself for being stuck at home alone
with Mom. Partners in life, my brother and I were like best friends.

In one hand, I grasped the handle of my carry-on baggage—an odd
little "ditty bag" that I had recently bought. It held the rudiments of
my first shaving kit, my Instamatic camera, pen and paper, and a few
reminders of life at home. I had bought the bag with my own money at a
military surplus store while shopping for my trip. I painted it red, white,
and blue in a stars-and-stripes pattern over its army-green canvas, in a

14

style currently in fashion with the anti–Vietnam War movement, though I was still too young to understand the war.

The line moved slowly to the ticket counter. Finally, my turn came. I wrestled with my sea bag to check it in, hoisting it up onto the scale. Mom paid for my ticket and we marched out to the departure gate.

Though I had traveled by jet plane only once before, I knew every corner of Sea-Tac Airport. Along with my neighborhood buddies, I had spent hours of idle time here, just hanging out, roaming the concourses in a pack. As travelers exited their planes, intriguing smells wafted out of the jet's cabin and into the terminal. Sometimes I detected hints of a mysterious, exotic spice that swirled within the scent of weary passengers. Airports fascinated me—they held secrets from faraway places.

My girlfriend, Patti, was not with me at the airport this morning. Together, we had decided to avoid a painful farewell at the airport—in public and in front of my mother. We had said our goodbyes the night before, lying in the grass along the shore of Lake Meridian, embracing while we watched explosions of 4th of July fireworks. We promised to write each other. I missed her already.

A voice from the PA system announced the boarding of my flight. I spun around to my family and shrugged my shoulders. When my mom grabbed me for a long hug, I faked dodging her kiss on my cheek.

"You . . . ," she said solemnly, pausing, "you . . . be careful."

I said goodbye to my brother, turned, and entered the boarding tunnel to the Western Airlines Boeing 727 sitting on the tarmac. After presenting my ticket to the attendant, I found my seat. I expected to see Uncle Jack but never saw him board the plane.

The plane accelerated to take off, climbed out, and punched through a low cloud layer into brilliant blue sky.

Mostly, I kept to myself for the flight, eating peanuts and wolfing down the breakfast. From my airline ticket, I saw that it would take over three hours to reach Anchorage. In Anchorage, I would change planes and fly for another hour to Kodiak Island. I attempted to read my book but instead stared restlessly out the window. Loneliness gripped me.

I had never before fended for myself. Locked within the confines of my seat, my body became rigid with tension. The distance from home grew with each second. Kodiak would be strange and new. I calmed down some, reminding myself that my destination—the schooner *Grant*—was very familiar to me.

I knew the decks of Jack's fishing boat well and pictured them in my mind. For as long as I could remember, my mom, my brother, and I had made special trips each year down to the docks in Seattle, to greet my uncles' fishing boats when they arrived from Alaska. Sinking into my seat on the plane, I occupied myself with these memories, relaxing a bit.

Up to the 1970s, the halibut schooners delivered fresh fish caught in Alaska directly to the wharves and warehouses of downtown Seattle. When we visited the boats, the scene was festive, a harvest of the sea celebration. The *Grant* floated below, tied to long pilings that plunged into the deep water of Elliot Bay. My brother and I would climb down to the boat and spend the afternoon clambering up and down masts and rigging like the monkeys that we pretended to be. Hovering above the boat on the dock, a bustle of women—wives and girlfriends—always milled around the crane hoist and unloading platform. I remember the contrast in smells of perfume and fish slime.

The unloading platform—an enormous, raised wooden table—sat on the dock next to the warehouse, elevated like a stage. Two crewmen—the "headers"—stood ready on the platform, towering over their audience. Their muscles bulged under their T-shirts, pumped up from working hard. They wore gloves on their hands and rubber boots on their feet, and over their legs and torsos were oilskin coveralls with a rope for a belt, cinched narrow about the waist. To me, their bodies looked like those of Greek heroes.

In one hand, the headers brandished long, heavy knives like machetes, raised and ready; in their other hand they grasped the T-shaped handle of a steel "gaff" hook. The J-shaped hook, the business end, snagged the fish.

On board the *Grant*, two more crewmen worked, hidden from view, loading the nets from deep within the fishhold.

About every five minutes, the dock crane strained to lift a net so swollen with fish that it hung like a huge teardrop. The crane operator raised the net up from the boat, swung it over the dock, and lowered it onto the table. The headers unhooked the net, and big fish covered with slime spilled out to the corners of the table. The two headers went to work, plunging their gaff hooks and slashing with the big knives.

They were swift and sure. With the gaff hooks, they lifted up the heads of the fish. With the long knives, they decapitated the fish with a couple of short, determined strokes. The headers worked furiously, eager to be finished with the fishing trip and heading for the pleasures of home. They raced to keep up with the procession of bulging nets that rose from the schooner. The headers slid the beheaded fish to other workers who sorted them into bins, or "totes," according to size and grade. Sometimes when my brother and I ran around up on the dock, we were hit in the back by rubbery halibut eyeballs—the size of golf balls—thrown by the idled headers. With glee, we hoarded the slimy orbs. When the headers bent over, occupied with another net full of fish, we shot the eyeballs back, giving them a barrage in return.

After heading, the responsibility for the fish was transferred to the crew of the cold-storage processing plant. The plant's crew cleaned and washed the fish with water and laid them flat on tall, wheeled racks with multiple shelves of sheet metal stacked seven feet high. The racks were either rolled on wheels into the cold-storage blast freezers or lifted by forklift trucks and driven directly into the freezers—smoking tunnels of ice. That's the last my brother and I saw of the fish—end of story. As a child, I had seen only the last hours of a fishing trip that had begun three weeks earlier.

How did a fishing trip begin? I hadn't a clue.

As the jet flew toward Alaska, I began to wonder what preparations were made for a fishing trip. How much work went into getting the boat ready? How did the gear function? It would take enormous work to load the boat with fish and deliver the catch back to the docks. How would the crew treat me? I fidgeted in my seat and stared out the window.

As part of my preparation for the trip, I had studied a map of my route to Alaska. On this day, a cloud layer obscured my view of the ground below. I had hoped to see the coastline and a marine corridor that I had heard about from my uncles—the Inside Passage, a saltwater maze that snakes through hundreds of islands that hug the Pacific shoreline. The seamless blanket of white continued while we passed over British Columbia and Southeast Alaska, hiding the peaks and glaciers of the Fairweather and Chugach mountain ranges. Three hours into the flight, the plane began its descent.

Landing in Anchorage, I knew that I had an hour on the ground until my flight left for Kodiak. I exited the plane, eager to explore a new airport. I veered around a corner in the main terminal and was startled to encounter a huge bear reared up on its back legs. Towering over me, it reached out with giant paws tipped with rows of scythe-like claws. Its fleshy red lips were curled back to reveal fangs inside its gaping maw, frozen mid-snarl.

It was very dead—and in a glass case. A big, brass plaque inside the case read, *Kodiak Brown Bear*. The plaque also flaunted the bear's vital statistics and the name of its executioner. I said to myself, "Oh great . . . KODIAK . . . that's where I'm going."

I laughed, knowing that I had only packed mosquito repellent. I had nothing in my sea bag to protect me from this beast.

For the rest of the hour, I combed through shops, checking out racks of magazines and shelves lined with Eskimo trinkets. The time drew short, so I headed for my departure gate, far removed from the central area of the terminal. Down a quiet hallway, I spied Uncle Jack. He flipped through the pages of a magazine, sitting in an empty row of chairs with his body folded into a seat that disguised his wiry and athletic frame— just shy of six feet, same as me. As usual, his short brown hair stuck out in places, seeming like a cowlick but always showing up in different spots. Of my three uncles, his face most resembled my mother's.

He looked up and said, "Well, if it isn't Dino," goading me right away. Jack enjoyed pushing people's buttons. I had acquired the nickname from Dino the Dinosaur on TV's *The Flintstones*. I tolerated people calling me this, though I didn't like it.

"How's it goin'?" My uncle's face broke into a wide grin, exposing the small gap in his front teeth.

"I'm doin' just fine, Uncle Jack."

Serious now, he asked, "Are you ready for this?"

"I guess so."

He nodded, making his cowlick bounce. Like his personality, his crown of unruly hair responded to its own discipline and order. He resumed scanning the pages of his magazine.

"Hey, wait a minute," I said, raising my voice. "How'd you get here?"

With a confused look, he lifted the brow of one eye, and said, "On a plane. How'd you think I got here?"

"Yeah, right. But, which plane? I didn't see you on my flight." I was sure that I'd seen everybody board the flight.

"Oh, I was there. You must have missed me."

Hearing him say this threw me off. It wasn't like me to miss anything.

For the last leg of our trip, Jack and I boarded a small twin-engine bush plane. On purpose, I chose a seat beside my uncle, next to a window.

Engines wailing, the plane raced down the runway. Through my window I saw the landing gear lift off. I was leaving the continent to land on an island. I felt anxious. At the same time, I felt peaceful, though I'm not sure why. I knew only one thing about my destination—it was the home of the giant brown bear.

5

KODIAK

THE PLANE DESCENDED AND I CAUGHT MY FIRST GLIMPSE OF
Kodiak Island. In clear skies, wisps of clouds milled around mountains
packed close to one another, topped by peaks that bore epaulets of snow
into summer. I glimpsed the small houses of a town, wrapping round the
base of a steep slope curbed by the sea. A row of docks and buildings
stuck like barnacles to the shoreline where black rock plunged into the
bay.

Bright flecks of color dotted the shore, looking like chips of paint—
primarily blue, green, and red—the colors of Kodiak's fleet of fishing
boats. I noticed that the green water of the harbor was marbled with a
pink-colored swirl, looking like whorls of an antique bookend. The curious patina meandered in a current leading out of the bay.

Where was the *Grant*, I wondered.

The plane wheeled in a turn, coming around a wooly buttress of
clouds. The orderly grid of a boat harbor came into view—perhaps
more than 200 fishing boats. Beyond the boat harbor, a row of industrial
buildings stretched along the shore, rimmed by still more fishing boats.
Next to the boat harbor, an oceangoing freighter stood tall, situated on
the waterfront in the middle of town. Looking closer, I could see that the

ship wasn't actually docked. The ship's hull was recessed to fit into the shoreline, with the earth rounding both the bow and the stern. A series of giant holes were cut into the side of the ship facing the water, gaping open to a narrow dock that serviced three fishing boats.

The plane's wings leveled off and I stole a look out the far side. The series of windows framed an archipelago of islands before an endless reach of shimmering ocean.

The plane's wheels barked, touching down on the runway. The pilot hit the brakes hard—too hard, I thought—and slowed the plane quickly, veering toward a small air terminal. Looking out my window, I saw that the runway ended in a rock cliff, rising up to a triangular peak thrusting over 1,000 feet into the sky. For me, the cliff face became a sobering first impression of Kodiak Island. I understood why the pilot had braked so hard now. After making a commitment to land in Kodiak, you don't get a second chance. I shuddered at the thought of an accident.

I grabbed my baggage in the terminal and searched for my uncle. I found him leaning out the door of a Checker cab, waving for me to come. Jack disappeared into the cab, ducking through one of its rusted-out door frames. I threw my sea bag into the voluminous trunk and dove inside the cab to find a seat.

The interior of the cab stank like cigarette butts marinated in beer. A layer of gray dust coated every surface. From the rear of the cab, all I could see of the driver was a huge mop of kinky hair. A baseball cap looked ready to spring off his head at any moment. I cranked open my window and a faint scent of vegetation mingled with the stench.

"Small boat harbor," my uncle instructed the driver, who stared ahead, uttering nothing in response.

The cab lurched and accelerated. In seconds, we hurtled down a gravel road, a tornado of dust in pursuit. We whizzed past skeletons of abandoned buildings and departed the vicinity of the airport, crossing a threshold into wilderness.

Alien to me, vegetation with huge leaves took over the scene, with foliage that seemed better suited for a tropical climate. Green patches of shrubs draped the mountains' flanks, bordered by legions of purple flowers, standing erect. At the base of the mountains, meadows of tall flowers with pink frills surrounded stands of stunted conifer trees. It looked like a vast garden, but a musty fragrance hung in the air, smelling like a bog on a warm day.

The road weaved along a hillside then cut through a break in the rock. A spectacular panorama expanded before me, offering ocean views of a group of islands, boats in the bay, and the town drawing near.

Down a steep slope below the road, I saw a cove. Adjacent to the cove sat some buildings and a silver behemoth of a boat looking like a Buck Rogers spaceship. Same as the big ship beached in the middle of town, this boat was also wedged into the earth. Odd, I thought.

I'd seen this strange boat years before, in my childhood. It had been a ferryboat serving the Seattle waterfront. The *Kalakala* looked more bizarre here.

To my right a row of wharves stretched along the shoreline for the rest of the way into town. Lined up against the hillside to my left were tall stacks of rusty steel cages, idle pots for catching king crab.

Clouds of steam billowed out of the fish plants. I stuck my head out of the cab. My hair whipped wildly and I caught the pungent aroma of seafood. The smell of fish began to permeate me.

Kodiak was home to more than a dozen fish processing companies. Each company's buildings were painted a signature color, a shade of gray, dirty white, green, or blue. Huge signs identified each complex. "King Crab, Inc.," "Peter Pan Seafoods," "Alaska Processing Company," and "Ursine Seafoods," to name a few. Central to the waterfront stood a group of warehouses painted a garish, bright sea-green. A big sign branded one wall of the largest building in the cluster: "B&B Fisheries." We rumbled along the waterfront in the cab, jolted in our seats by deep chuckholes.

A few seconds before entering the town, the gravel road turned to asphalt. At the base of the mountain looming over the town, I saw little houses painted in drab colors, or with no paint at all. Each had small paned windows and a steep roof. The cab slowed and became quiet.

On the left, we passed a small parking lot surrounded by brown buildings that were nothing but big wooden boxes. From reading the signs, I saw they were taverns and liquor stores, a hardware store, and a drug store. On the right, a small building with a flagpole outside marked the harbormaster office. The clamor of rock resumed under the cab's tires and I realized that, in seconds, we had passed through the center of town.

The road weaved to parallel the shore, taking us past the big beached ship I had seen from the airplane. I gaped at a wall of steel as it passed by, seeing the ship's name painted on its stern—*Star of Kodiak*.

I spoke up and asked Jack, "What's with the ships stuck on land?"

"A tidal wave from a few years ago destroyed all of the buildings along the water in Kodiak," he replied. "To get businesses back on their feet, some boats were grounded in place to become prefab buildings."

I had to catch myself as the driver slammed on the brakes, making the wheels dig deep into the gravel. A cloud of dust swallowed the cab. We had stopped in front of a gangway that led down to a floating pier. A ragtag flotilla of fishing boats in all shapes and sizes lay alongside, decorating the shore of the channel below. I emerged from the cab, blinking grit from my watering eyes.

Looking down to the docks, I saw that the boats were rafted out, three deep on each side—demand exceeded the supply. I recognized one of the larger boats as Uncle Jack's schooner, the *Grant*, tied outside of a scruffy seine boat. Compared to the seiner, she was regal.

Jack paid the driver. The cab rolled away.

"TRANSIENT VESSEL DOCK" read a small sign next to the gangway. Transient vessels? What's that about, I wondered. I considered transients to be homeless people—alcoholics.

I heaved my sea bag awkwardly across my neck and shoulders. The bag threatened to crush me. I felt self-conscious, thinking I must look like an ant carrying an impossibly large burden. I followed Jack down the gangplank, staggering.

Still wrestling with my load, I crawled over the railing of the seine boat and made my way across her deck, which was strewn with fishing gear, buckets, and brushes. I threw my sea bag over the *Grant's* rail and swung myself on board. My feet landed with a thud on the *Grant's* heavy wooden deck. My heart pounded. A breeze of a perfect afternoon washed across the deck.

Jack stood waiting on deck. "Welcome aboard," he said in mock formality, offering his hand and bowing slightly.

Jack loved his sarcasm. That I knew. He sensed my nervousness and was toying with me. I surveyed my new surroundings and fought to contain myself—and to ignore Jack.

The boat reeked of fish, yet I could see no fish, slime, or blood. I halted to gawk at one of my fellow crewmen, a man who looked to be around sixty, working on deck with piles of rope stretched in front of him. A wisp of gray hair encircled his bald head. A network of bulging muscles rippled up and down his arms. The fitness of the man astounded me. I was a young athlete in my own right, but compared to him I had sticks for arms. I clearly lacked the proper equipment for my new job.

He was short in stature, barely five foot six, and he embodied an aura of confidence. His bright eyes shone deep within sockets divided by a bulbous nose. A white undershirt with the sleeves ripped off exaggerated the breadth of his square shoulders. He smiled as he moved his hands, listening to Jack and observing my reactions. He was connecting the ends of two ropes, using a brass spike to interweave their strands. He set the tool down and reached out a thick hand to shake mine, his demeanor easy and slow.

"Hi there, kiddo. You must be Dean. I'm Kaare."

"Nice to meet you, Kaare."

"The aft cabin's full. There's two of us down there already." Motioning toward the bow with his hand, he said, "Go ahead and stow your sea bag down below in the fo'c'sle, and take your pick of the empty bunks."

I liked Kaare, and I got the feeling that he liked me. I hoped so. I knew that I needed to get along with these people. I also wanted to get settled into my new home, so I headed toward the fo'c'sle.

At the start of the bow deck, the companionway gaped open like half of a rectangular clamshell. Its shape served as a shelter for the entrance to the fo'c'sle (short for "forecastle"), nestled deep within the bow. I leaned into the opening of the companionway and dropped my sea bag to the floor below, then I descended into the fo'c'sle, eight feet down the ladder.

Though it was daylight above, the fo'c'sle languished in the heat and darkness of equatorial midnight. This small space functioned as a bedroom for three and a kitchen and dining room for six. Aromas lingered in the stagnant air—men, food, tobacco, and fish. I wrinkled my nose, recognizing the vapors of diesel and pine tar. I had learned about pine tar—a wood preservative made from tree sap—from watching my uncles slather the pungent oil onto the lumber of their boats. In the far corner of the fo'c'sle, a large diesel stove growled with a quiet roar. The black metal of the stovetop radiated heat, casting invisible waves that tingled my cheeks and brow.

Except for the galley stove, everything about the fo'c'sle was compact and cramped. Only the aft section, a space three feet wide and eight feet across, afforded enough headroom to stand up straight. At home, living in the Cabin, my brother and I shared a bedroom one foot smaller each way than the whole fo'c'sle.

The corner of the stove abutted a narrow countertop with a ship's kitchen sink. There was no sink faucet. Instead, a cast-iron pump towered over to one side. Pumping the handle hard brought fresh water spilling out of the spigot into the sink. The pump looked antique and out of place, better suited for watering farm animals.

The sink was small in diameter but was two feet deep. I supposed that a shallower sink would have spilled its contents when the boat rolled from side to side.

Just forward of the sink, a couple of galley cupboards were built into the wall up to where the bunks started. Locking latches secured the cupboard doors.

A half dozen bunks—each six feet long and twenty-four inches wide—were cut like deep shelves into both sides of the bow. Less than three feet of headroom barred any chance of sitting up in these beds. One bunk started forward of the cupboards, with storage below reserved for sea bags. Two pairs of lower and upper bunks covered most of the wall on the port side, with the head of the most forward bunks adjoining in the peak of the bow.

From the forward bunk, an occupant on one side could reach across and touch the nose of the person sleeping on the other side. I wanted more privacy than that, so I chose an empty upper bunk further aft, on the port side of the galley next to the table. With a grunt, I heaved the sea bag into my bunk, nearly filling the space. It reminded me of a little niche where the bones of the dead are laid to rest—a coffin in a crypt.

The triangular galley table was wedged into the bow, between the bunks. The raised sides of the lower bunks doubled as seatbacks for people dining at the table. A wooden box hanging down from the ceiling was packed with condiments in glass jars.

I wasn't claustrophobic, but the oppressive heat in the fo'c'sle made me queasy. I figured I had better get out. I wanted to avoid seasickness at all costs. I loathed throwing up.

I rushed up the ladder into the blinding sunshine. Jack still lingered on deck, chatting with Kaare. The salt air refreshed me.

Kaare had spent his seven-day layover in Kodiak instead of flying home like Jack and the others in the crew. After completing a fishing trip, most boats in the halibut fleet followed a voluntary system of a week-long layover.

I noticed two tall, bearded, barrel-chested men strolling down the dock's gangplank. They climbed over railings and crossed the deck of the seine boat. For being such big men, their ease and agility surprised me. I recognized Chris, Jack's brother-in-law, and gave him a little wave.

Jack looked up. "Ah . . . Chris and Wally."

Chris had a thick ponytail of red-blonde hair that flopped down his back, counterpart to his long, scraggly beard. He wore heavy, black-framed glasses like Roy Orbison's. I had met Chris at holiday get-togethers at Jack's house, and I liked him. He was eighteen years old and seemed wise and mature to me. All my life, I had deferred to the adage, "respect thy elders." Chris's beard added to his authority—especially since I couldn't yet grow one.

Wally was older and a little heavier than Chris and looked powerful. He had a Santa Claus twinkle in his eye and a jovial smile. His hair and beard were short and dark and I stared at the stubble that covered his face in bristles packed so tight I could barely see his skin.

I dawdled, hiding behind Jack's shoulder while Wally met Jack with a handshake. Kaare continued to work on his gear.

"I'm good, Jack. And you?"

"Couldn't be better. Good layover?"

"Dandy. Played with my kids mostly. The time at home was too damned short though . . . as usual."

Chris nodded in agreement. Kaare kept working.

"Well . . . here we go again." Jack paused, noticing Wally check me out. "Oh yeah. This is Dean. Dean meet Wally."

We shook. His thick hand crushed mine.

"You already know Chris."

"Hi, Chris," I said, and reached out, expecting more abuse of my hand. Chris shook my hand gently and smiled.

"You guys got your gear work done?"

"Yeah . . ."

"You bet."

"The bars got the best of me, Jack," Kaare joked. "Don't you worry. I'll get mine done before the sun goes down."

Jack grinned back at Kaare. The summer days in Alaska are long—that I knew. Considering the tall piles of rope stacked around Kaare, I wondered if he had cut it too close.

Jack's face became grim. "I hope the fuckin' weather is better on this trip

". . . better than the last." He cursed, sounding matter of fact. Still, I flinched a little hearing him swear. Jack had never cursed around me before.

"I've made an appointment with B&B for getting ice and bait tomorrow morning."

"How much bait?" Wally asked the questions, making me wonder if he was the leader of the crew.

"I think that I asked them to get us . . . let's see . . . uh . . . 3,000 pounds of salmon, 5,000 of herring, and 2,000 each of cod and octopus."

Twelve-thousand pounds. Of bait? I was shocked.

"Any particular way you want it arranged in the hold?"

"Nope. Whatever you've been doin' is just fine with me."

"I'll talk to Freddy."

Who is Freddy, I wondered, then realized that one person was missing. Only five of us were present on deck, counting Jack.

"Good idea. Okay, now . . . grub. We won't have to scrounge up a truck to get the groceries. Kraft's is sending their van from the store. It's coming around noon."

The crew nodded in approval.

I heard an airplane engine rev up. A floatplane appeared down the channel, coming from the direction of the *Star of Kodiak*. The narrow channel concentrated the noise of the airplane as it took off, the volume building to a point that forced me to plug my ears. The plane whizzed along the water, passing and lifting off so close by that I saw water streaming from the floats. I loved it. I had inherited my father's fascination with planes.

When the plane's roar had subsided, the conversation on the *Grant* continued.

"How 'bout ice? You want the usual, Jack? About ten tons?"

"Yeah. Let's see how it comes on board and have the guys on the dock cut it off when we're getting close to full."

"No problem."

From the look on his face, my uncle was scanning a mental checklist. "Okay," he pondered, asking, "We're topped off with fuel and water, right?"

"Yup."

"How 'bout hooks and gangions?"

"No sweat. Our stock is good," said Wally. "I brought up a few more with me on the plane."

I knew what gangions were and that they served the same purpose here as the fishing leaders I used for catching trout—a means for connecting a hook to a larger, primary fishing line.

"Good. I guess we're all set then. We'll start in the morning at eight o'clock."

For my uncle and me, a new journey began that day. I was familiar with the uncle who showed up with his wife and two daughters at Gramma's house for Christmas and Thanksgiving, wearing a plaid shirt and corduroy pants. I knew the holiday version, clean-shaven and well-mannered. I knew the exterior of my uncle, not the entire man.

"Go ahead guys, the night is yours," said Jack, dismissing the crew. "Chris . . . you keep an eye on Dean."

Unsure of what to do, I looked to Chris for direction. An impish grin twitched his beard. He waved for me to follow him. "Let's find Freddy and go uptown. He's on board the *Lindy* . . . just down the dock."

6

FREDDY

ON OUR SHORT STROLL DOWN THE DOCK, CHRIS TOLD ME about the crew of the *Lindy*, giving me my first account of fishing lore. He said that this crew was infamous for hosting parties that plowed full speed through nights and days. Chris told me that Kaare, who was Freddy's stepfather, was a new hire on the *Grant*. Freddy had gone to see the *Lindy*'s crew to follow up with them about a job swap involving Kaare. Chris was sure that the merrymaking on board the *Lindy* had sidetracked Freddy.

Climbing down the ladder into the *Lindy*'s fo'c'sle, I descended into a party rich with libations and revelers. A platoon of beer bottles covered the table. Chris pointed out Freddy, who was in the midst of a slurred exchange, howling about some misadventure in the bars uptown.

At about five-foot-six-inches tall, counting the heels of his black leather Beatle boots, Freddy looked like a compact version of a teen heartthrob, a cross between Elvis and Bobby Darin. A thick shock of greased black hair flipped back off his forehead in a wave, tapering to a sharp point. He wore a white T-shirt and black denim jeans that fit tight over his muscles, stretching the fabric. He sported an irresistible smile. Freddy had "fun" written all over him.

My eyes widened when I noticed an open glass jar sitting on the galley table, half-filled with crumbled green leaves—marijuana. A couple of packs of rolling papers sat on top. Pot smoke lingered in the air. In the last year, I had become familiar with pot's aroma as well as its effects, experimenting with hits off skinny little joints and coughing my guts out while hiding in the woods behind the Cabin.

Chris jerked on Freddy's wrist to get his attention. Tearing himself away from a story that I couldn't understand, Freddy bid his hosts farewell, gasping for breath from laughing, acknowledging me with, "Hey, kid." The three of us left the *Lindy* and walked uptown at a brisk pace, arriving in a couple of minutes at an enclave of taverns and bars organized around three sides of an open square that faced the boat harbor.

Freddy and Chris turned and disappeared through a door into a bar marked "Tony's."

Before I knew it, I was in a dark room, plopped onto a barstool, sitting next to Freddy and Chris. My attention swiftly riveted onto a woman dancing on a small stage, buck naked except for spike-heeled shoes. She writhed in the flash of a strobe light with her back arched. A pained look distorted her features. She would have been pretty, but it looked to me like she was being jolted with electricity. She thrashed to the beat of muffled rock music blaring out of a jukebox with blown speakers. I couldn't take my eyes off the dancer, but I soon quit my obsession when I discovered other strippers strutting around in the buff.

Whether my mother had this in mind when she sent me north to Alaska, I didn't care. As far as I was concerned, my trip had developed into an adventure with unspeakable benefits.

An older woman bartender walked over, blocking my view of the stage. I tried to look around her but stopped, realizing that she was glaring directly at me. I knew that the drinking age in Alaska was eighteen. I guessed that she had caught sight of my peach-fuzz sideburns and fresh rosy cheeks, not to mention the slack-jawed, dumbfounded look on my face. It was time for me to get kicked out.

She cocked her head to scowl at Freddy, then cocked her head back at me. An ash broke off from the cigarette stuck between her lips. She rolled her eyes over to Freddy, gave him a harsh look, and said, "Why the fuck did ya bring dis kid in here?" I tried my best to look invisible.

Freddy responded with a Hollywood smile, lowered his eyes, and purred to the woman, "Give him a beer. Okay, sweetie?" She furrowed her brow and paused.

Not breaking her gaze with Freddy, she reached below into a cooler recessed into the counter and pulled out a bottle of beer. With an opener, she ripped off the bottle cap and slammed the bottle on the bar with a loud crack. I was amazed that the bottle didn't shatter. Foam spewed out the top.

Still looking at Fred, her scowl evaporated and a smile spread over her face, revealing rows of stained teeth. In a fuzzy, whiskey voice she said, "How ya doin', honey? I've missed ya."

To me, she was intimidating as hell itself—a bleached-blonde dragon lady with big hair and fake eyelashes. Her blouse squeezed wrinkled cleavage out of a neckline that drooped far beyond reason. Not only did she scare me, but Freddy knew her. I had never known anyone who knew this kind of woman.

From her body language, I sensed that she controlled this joint. Somehow I knew that with just a simple nod in my direction and a little wave of her hand, a bouncer would appear to throw me out. I could imagine the bouncer hauling me to the door by the scruff of my neck and launching me headlong into the alley. Instead, the dragon lady bestowed generosity and allowed me to hang out on a bar stool and drink a beer. She turned to help another patron.

"Whaddaya think, kid?" said Freddy. "Pretty nice place, huh?"

Chris just chuckled.

I didn't know what to think. I had never been inside a strip club. It astounded me—a macabre scene of strobe lights and undulating flesh. My eyes adjusted to the darkness and I saw more. Advertisements from beer companies decorated the paneled walls—a couple of mirrors and a half-dozen posters—with fist-sized holes between them, punched clean through.

I explored the shape of my beer bottle with my hands and withdrew, trying to gain an understanding of burlesque. Along with the sights, the alcohol in my blood seduced me. I sat and watched, afraid to speak.

* * * *

Hours passed and I became exhausted. I could see that my endurance paled in comparison to my fellow crewmen. They were more alive than ever—hooting it up, buying table-dances, chatting up the dancers, and swapping bullshit stories with fishermen from other boats. A fight broke out between a man and a woman—punches were thrown, sometimes missing the target by an arm's length. The drunken performance would have been funny if it hadn't been real.

I had had enough. Alone, I shuffled back to the *Grant* before midnight. I was amazed that the northern sky was still light. Shouldn't it be pitch black by now?

Back at last in the *Grant*'s fo'c'sle, I was relieved to find it cooler. I stripped down to my T-shirt and underwear and crawled up into my bunk.

Sleep evaded me as my mind retraced my path from the Cabin here to Kodiak—a journey from Seattle to Alaska that had delivered me to a strip club. My mind gyrated with fantastic images, ranging from landscapes to bodyscapes from the plane and from the bar. I lay awake, recalling scenes from the bar's dark corners. Far from the brightly lit stage, dancers performed at tables that intimated privacy. Female nudity was an exciting novelty, but the relationship between the dancers and a few of the bar's customers disturbed me. The men seemed agitated and angry as the dancers thrust temptation smack-dab in their faces. I saw too many people who had drunk themselves into a stupor, too many who were hideously drunk and delirious.

I shuddered in my bunk, recalling two men who were waxing philosophic to each other spit ejecting from their lips as they blithered, taking turns pounding fists on tables, dead serious, hammering home a point. What were they trying to communicate? They spoke in tongues. I was pretty sure they were speaking English, but I couldn't understand a word.

Until this point in my life, I had been exposed to the TV version of a drunk—a Red Skelton in makeup and costume, pretending to be a bum, a happy drunk, having fun. The people in Tony's weren't comedians. That night, I saw the dark side of being drunk.

7

THROWING
THE LINES

I STIRRED FROM HEARING A SHOUT AND WOKE UP, TRAPPED, sandwiched by the sides of my narrow bunk. Disoriented, I peered out from behind the towel that I had strung up across the opening to my bunk the day before, copying how Chris and Wally gained some privacy from the rest of the galley.

I craned my head to look to the top of the fo'c'sle ladder in the direction of another shout. The square shape of the companionway framed electric-blue sky, blinding me. I raised my hand to deflect the brilliance and cursed the pain I felt in my head. I glimpsed Freddy above on deck, dashing by. From my vantage in an upper bunk, I looked down and around the darkened fo'c'sle. The crew had vacated their bunks. Why didn't they wake me? I checked my watch—eight o'clock in the morning. I had better get up there. Something was happening.

I rolled out of my bunk and climbed down to the fo'c'sle deck, using several handholds and the side rail of the bunk below as a step. I snatched on my pants and shirt, slipped on my shoes, and raced up the fo'c'sle ladder in a sprint.

The *Grant* was no longer tied to the transient dock. The boat moved slowly, emerging from a channel into a large bay in front of town. The air was fresh and marine. The sun's rays warmed the shirt on my back. Through my puffy, sleepy eyes, I surveyed Kodiak—a breathtaking view of a fishing community set within nature. Boats surrounded me. Vessels mated with docks that serviced the boats, coming and going. From here on the water, it was easy to see that this town existed through fishing.

To my right, a forest of masts poked above the boulders of a seawall protecting the inner harbor. Ahead, fishing boats nestled against docks of several fish plants. A breeze fluttered a crowd of sea gulls. Clouds of steam spewed from buildings and swirled across a backdrop of mountainous shoreline. Picture postcards without frames surrounded me in every direction.

"Hey, Chris. What's going on?"

"We're moving the boat over to B&B to get ice and bait."

I paused and took in a deep breath. In my head, I repeated, "Ice and bait . . . We're getting ice and bait!" This was it—the real thing. Today, the *Grant* would come alive as a fully functional machine for fishing. I was one of the crew, filled with pride—and feeling a little unnerved.

The *Grant* slipped through the glassy water like a giant canoe. On deck, the sound of the boat's engine was undetectable, but my feet sensed the vibration. The boat's passage disturbed the water's surface to make a delicate V of a ripple emerging from each side of the bow. Jack steered from inside the pilothouse, using the power of momentum to coast ahead. From the time I appeared on deck, the boat's heading hadn't strayed, zeroed in to a point onshore lined with docks and warehouses. I pretended I rode a giant torpedo, locked in on a target.

A large building painted sea-green rose directly in our path. I made out the big sign for B&B Fisheries. The docks of other fish processing plants flanked B&B's facility, all perched on rows and columns of log pilings along the waterfront.

The *Grant*'s crew stood by, waiting and staring ahead as we approached. They stationed themselves in pairs—Chris and Wally on the bow, Freddy and Kaare on the stern. One held a heavy coil of thick rope, the other a broom. Close now, the salt air became tainted with the smell of rotten fish, which started first as a whiff, and then became more penetrating as we neared B&B's dock, now towering thirty feet above the deck of the *Grant*—almost as tall as her two masts.

Starting as a deep grumble, the engine began to roar. Just as I began to fear that the boat would ram the dock, Jack gave the engine a vigorous burst of power in reverse gear. Black smoke belched from the exhaust stack on top of the pilothouse and the deck rumbled under my feet. Beating the seawater around the stern into froth, the *Grant*'s propeller struggled to stop her forward progress. The deck rose slightly and tilted to port side. Jack twirled the *Grant*'s wooden-spoked wheel in the pilothouse to starboard—to the right—to move the rudder. The boat slowed and pivoted from a point near the bow to swing sideways and slide to lie alongside the dock. The crew flew into action.

When the boat came within range, Chris heaved the line to one side of a piling. Wally had extended the broomstick from behind the other side, allowing the thick line to flop on top. Wally pulled the rope back, now looped around the piling, and secured both ends to a massive cast-iron cleat bolted to the deck of the *Grant*. At the same time, Fred and Kaare tied the stern line. Jack flew out the pilothouse door and hustled down to the main deck. He grabbed another tie-up line lying ready, neatly coiled on top of the hatch cover, leaned over the rail and swung the rope around the piling, snatching the end as it flipped around. Like a rodeo calf-roper, he whipped the line around the steel horns of a cleat positioned amidships, making it the third of three lines holding the *Grant* fast to the dock. The pace of Jack and the crew relaxed now. The job of docking the *Grant* had taken the work of five people. This surprised me. I expected docking to be a simple process, taking two people, three at most.

Cool air wafted up from under the pier, laden heavily with a putrid, rotten smell. In the deepest reaches under the dock, I spied a rat crawling over slimy boulders. From the elevated floor of the processing plant adjacent to B&B, thick braids of water cascaded into the ocean through many drains. Hitting the water below, bubbles formed gobs of pink foam that came together to form rubbery rafts. Despite the delicacy of the rosy color, I guessed that the revolting stench and the rafts of foam swirling around the pilings and reaching out into the harbor were associated with one another.

Pointing, I asked Jack, "What's that?"

He jerked, his thoughts somewhere else. "Huh?"

"What's with the foam?"

In a lopsided grimace, my uncle chewed at the inside of his lip. He looked uptight. "They're processing shrimp," he said.

Without explanation, Jack spun around and reached for the steel rungs of the ladder that ascended to the dock. Climbing fast, he vanished over the top. A few minutes later, dockworkers appeared, craning their heads to peer down. I heard an electric motor start, the noise ascending to a penetrating wail that grated on my nerves. The dock's crane shuddered and began to move, sweeping its arm in a large arc. Fred, Kaare, Wally, and Chris took positions on the corners of the barndoor-size hatch cover and lifted hard, removing it from the hatch combing. They tilted the hatch cover up to lean against the boat's railing. The hold of the *Grant* gaped open, ready to take on supplies.

The dock crane lowered thousands of pounds of bait down to the *Grant*, loaded on pallets. From fishing for trout back home, I had learned that the choice of bait type and its quality were vital for catching fish. I checked out each type as it came aboard—frozen herring, salmon, octopus, and cod. Wally let me tear open a box of frozen octopus bait, revealing rows of suckers along five-foot tentacles.

I became that kid on the docks in Seattle, captivated by each new thing. I became distracted, and even though I knew I had to work now, I couldn't help myself. Everything was a novelty. Either I wanted to poke and prod, or watch and wonder. But this wasn't like one of my visits to the docks in Seattle. This time I was going fishing. As much as I tried to stay focused, the crew had to goad me into work.

"Hey, Dean. It's called a sea gull, for chrissakes." Back home around Puget Sound, I'd seen tens of thousands of sea gulls, if not millions.

"Dean, snap out of it. Get over here . . . and stop fuckin' around. You can help Wally take ice. Get on your skins! . . . For crying out loud, do something!"

Skins, short for oilskins, are the heavy-duty raingear of fishermen. I struggled getting my oilskins on until I figured out that I needed to pull the tops of my boots up first. I had put on my big rubber boots that day before going on deck, and I'd copied how the crew doubled down the tops of their tall, thigh-high boots, like pirates. Pulling them up, they extended to just below my groin. I soon found out why I needed my skins for taking on ice. Of all the steps in provisioning the boat, the best to watch, or better yet, participate in, was the process of taking on ice.

B&B's workers lowered one end of a giant plastic hose from the dock reaching down into the *Grant*'s hold. The hose was eight inches in diameter and at least fifty feet long. The other end of the hose disappeared over the lip of the dock out of view. I slipped over the edge of the hatch combing, hung by my arms, and dropped down into the cavernous fishhold with a thud, joining Wally. It smelled musty down there, like a root cellar.

The contour inside the hold followed the shape of the boat's hull, rounded from the bottom and up the sides. A vertical wall to forward, the bulkhead, separated the hold from the fo'c'sle, the aft bulkhead from the engine room. Along the hold's outermost surfaces, galvanized refrigeration pipes snaked back and forth in tight S-turn coils, surrounding the hold in cold, protecting the ice from melting. The ice was necessary to preserve our catch of fish.

From up on the deck, Chris hailed one of the workers on the dock to start the ice hose, "Go ahead. We're ready." Chris peered down at me and warned, "Watch yourself."

Quiet at first, but slowly gaining in volume, a sound like an air-raid siren howled from the end of the hose as a powerful blower from far inside the fish plant spun up and gained speed. In a few seconds, the huge hose gave a powerful jerk and an eruption of crushed ice started spewing out of the hose. The hose convulsed like a mammoth snake, taking the strength of both Wally and me to keep it under control. I was having a blast.

In fewer than fifteen minutes, the four center sections of the hold had been filled with 20,000 pounds of ice. Wally yelled up to Chris, "That's it. Shut her down." Chris hollered, relaying the message to the dock. The howl of the hose subsided and became silent. I looked at Wally and saw that both of us were coated from head to toe in ice, flocked with a white crust like a Christmas tree. I wished that someone from home could see me like this.

Next, the crane lowered five pallets of food to the deck. I'd never seen so much food outside of a grocery store, enough to stock the boat for a three-week-long fishing trip.

Three weeks. . . . Time and again I'd been told how long we'd be gone on this trip. Most recently, Chris reminded me while we provisioned the boat. Everybody on the crew seemed to want to stress this point. To me, two days at sea sounded like a long time. I wondered how three weeks would feel.

"We'll travel for two days or so," said Chris, "until we get to where Jack wants to start the trip."

"Why three weeks? Why not one or two?"

"That's as long as halibut will stay fresh packed in ice, and we want to make the most out of each trip. We'll work seven days a week, eighteen or nineteen hours a day. Then, after three weeks of fishing, we'll head back to town and deliver the fish. Considering how the fishing's been these days, it's unlikely that we'll fill the boat before then."

Jack and the crew started their fishing season in April and had completed three trips. This coming trip would be their fourth. The year had been going well—nothing remarkable, just a series of average deliveries.

"Will we come back to Kodiak?"

"We're never sure which town or which buyer we'll deliver to. It depends on who's paying the best price. Kodiak's been very competitive this year, and it's the closest town of any size to where we're going to fish. Seattle's always got the best price, but it takes a week each way to get there and back—that's traveling nonstop. If we sell in Seattle, we can't catch as much fish because we can't fish as long. We'd have to quit a week early because of the extra running time. Most likely, Kodiak will be the next town we see."

"I hope so," I replied. I liked Kodiak.

Jack came up from the engine room. He looked flushed and hot. Grease smeared his arms and a black stripe crossed his forehead where I guessed he had tried to wipe sweat so it wouldn't drip into his eyes. "Okay, guys. The boat's squared away. Do any of you have anything else to take care of while we're still in town?"

A couple of head shakes and a few grunts indicated that they didn't.

His chest swelled, taking in a deep breath. Raising his voice some, he announced, "Well, let's throw the lines and get goin'." I couldn't tell if it was apprehension or excitement I heard when he spoke louder, or both.

The crew marched to their various positions and let the lines go from around the pilings. I watched as the crew coiled the tie-up lines and then stowed them.

From up on the dock, workers hung around to watch while Jack maneuvered the *Grant* away from the processing plant. Feelings of exhilaration and fear jolted through me—simple and unsophisticated fear. I was scared, frightened in a good way. I sensed that the workers up on the dock inspected us with a degree of skepticism, like they might not see us again. I knew that fishing was risky. Was it worth it? I didn't know.

The *Grant* pulled away from the dock, slowly moving forward by the propeller, an invisible force below. To the west, the sun reflected a liquid trail of gold toward the open ocean. Jack and the crew paused, taking a moment to look back to the shore, seeming to pay homage to terra firma as the island panorama reeled by and grew smaller as the distance from shore increased. The ocean began to dominate our surroundings.

The bow veered, and we passed close to a red buoy marking the harbor entrance. The deck leaned over—heeled—to starboard, making me stumble to catch my balance. In the same moment, the entire lumbering mass of the *Grant* heaved up and I felt the first sea swell of my life. I could barely see the very long and rounded wave. The swell was slight and smooth and indistinct, not much more than a foot high, rising and falling like a bed sheet.

That the boat moved at all startled me. In Seattle, the *Grant* always felt permanent and unmoving. I didn't expect this kind of movement. My uncles' boats—the halibut schooners—were the biggest and most massive boats that I knew. As a child, I worshiped their integrity. Ignorant, I believed that the huge bulk of the *Grant* would remain stationary here—not allowing the sea to toy with it. In the dips and rolls that followed, I was dumbfounded to realize that this large boat was trivial in relation to the ocean. I had had complete faith in the *Grant*, until now.

Rising from the recessed pulpit of the companionway, Freddy hailed out, "GRUB!" He marched across the deck with a haughty look on his face, opened the pilothouse door, went in, and slammed the door shut. Jack came out shortly thereafter. I followed my uncle, climbing down the ladder into the galley.

The boat rose and fell with each plunge into the waves, making the heavy planks inside the fo'c'sle creak and groan with pops, growls, and other noises of complaint. Glass jars clinked as they slid in the condiment box above the table. Chris reached up and jammed a wadded up towel inside the box to stop the noise.

I dished up a plate, but soon quit eating and just poked at my meal. I tried to convince myself that my lack of appetite was a sign of nervousness and not seasickness. I rarely lost my appetite, except when sick. In a variation on "the bigger they come, the harder they fall," I fabricated a new theory on the spot:

The less eaten, the less vomited.

I didn't realize then that seasickness doesn't care whether a stomach's empty or full. No matter what, two convulsions and it's voided—and then you just keep on barfing.

I needed some fresh air at once. I decided to get out of the fo'c'sle. I'd go visit Freddy up in the pilothouse.

I surfaced on deck and marveled at seeing the *Grant's* tall masts slicing through the sky in great arcs, swaying gently from side to side against a backdrop of blue sky. Nearing the rail, I looked ahead and saw the bow cut a giant furrow in the water like a farm plow, turning over a boiling wave. The boat made a hard roll to port, surprising me once again in the hull's ability to move, and its speed in doing so. Thrown off balance again, I grabbed at the railing to catch myself. Seawater gushed across the deck through a row of holes in the side of the boat, the scuppers. These holes allow water to run off the deck. I knew about scuppers but had never seen them function. As the water slowly cleared from the deck, I jumped away, trying to keep the rolled-down tops of my boots from getting wet. I ran to the pilothouse to join Freddy.

The two of us crowded the tiny wheelhouse. He sat on the narrow pilothouse seat and stared blankly at the sea ahead. Our course paralleled the southeastern shore of Kodiak Island, about two miles offshore. From here, the features of the rugged mountains were diminished, replaced by waves that grew larger. Two of the four tiny windows in the pilothouse—two feet tall by one foot wide—had been lowered. The openings were too narrow for my shoulders; only my head made it out the window to feel the crisp salt air, much cooler since leaving town. I looked down at the waves right next to the boat.

"How big are they, Fred?"

"I dunno. . . . What are ya talking about?" he responded, coming out of a daydream.

"The waves. How tall are they?"

He looked outside and smiled.

"I suppose they're three . . . maybe four feet."

I registered four feet as the biggest swells that I had ever seen.

I studied the waves. They traveled along at the same speed as the boat, surfacing in an undisciplined mob. It was difficult to calculate their speed, because when any individual wave would materialize, it rose out of the sea, crested, and got gobbled up by its aggressive mates. Occasionally, a wave that had just disappeared would emerge defiantly from under the surface, having stolen energy from another, less viable wave, and make another futile attempt to exist.

A breeze followed us at about our own speed—nine knots—another way of saying nine nautical miles per hour—approximately the same as ten miles per hour registered by an auto's speedometer in the statute system. Years earlier, I'd learned about these terms from a nautical handbook that I'd found in Gramma's basement.

The rolling of the boat had made me sleepy and I felt worn out. "I'll see ya later, Fred. I'm goin' to bed." The day had bombarded me with incredible sights and new understandings.

"You be careful when you go forward," he warned. "You haven't got your sea legs yet. Make sure that you have your hands on something to help you when you move across the deck."

"Ha!" I thought. To me, it sounded like a dare.

Confident, I set out across the deck, determined to *avoid* using my hands. In the midst of my second or third step, the boat made an exaggerated movement under my feet. *Not* using my hands, I slammed against the rail with my ribs taking the impact. I tried to stifle a loud grunt. I hobbled the rest of the way, holding my side with one hand and groping the pipe rail with the other until I reached the companionway. I didn't look back as I climbed down into the fo'c'sle, certain that Freddy had watched me and now smirked from the pilothouse.

I burrowed into my bed. Lulled by the womblike rumble of the rocking boat, I succumbed to sleep quickly, held tight and cradled by the sides of my narrow bunk. It felt so good.

8

ALITAK

I AWOKE THE NEXT MORNING STARTLED BY THE SILENCE. The quiet resonated, like the emptiness that follows a gunshot. Apart from the purr of the stove fan . . . nothing. My attention focused. The boat lay hushed and unmoving. The *Grant* had stopped traveling.

Looking around, I saw that my partners in the fo'c'sle—Chris and Wally—had left their bunks. For the second morning in a row, I woke alone. My awareness expanded. Muted sounds bent through the crook of the companionway—a shout, the toot of a horn, the whining engine of a forklift truck, and, in the background, the white noise of cascading water. A fish processing plant, I guessed.

I rubbed at the sleep crystals in the corners of my eyes. My body felt restless, ready to get up. A few sore muscles ached to be stretched. I could tell that I had slept hard and for a long time.

My uncle hadn't mentioned making another docking—a landfall—before we started fishing. Neither had the crew. I was confused. I dressed and climbed up to the deck to check things out. My eyes hurt, adjusting to the light of day.

The *Grant* floated immobile, secured by lines to the big dock of a fish processing plant, a facility much like B&B's in Kodiak, cantilevered off a rocky shore on stilts of pilings. But here, instead of a town, wilderness encircled the domain of the plant. The dock hung over the green water of a small bay surrounded by green hills covered in grasses and spotted with flowers. A couple of stunted evergreens cowered in a notch at the base of a steep hill adjacent to the fish plant. Thick vegetation covered harsh outcroppings of shale—my first landscape of coastal tundra. Sometime during the night, the *Grant* had passed beyond the western extremity of forested land along Alaska's Pacific coast into an unfamiliar world to me—one without trees. Where I grew up in western Washington, the terms "wilderness" and "forest" were interchangeable.

Jack leaned against the rail, soaking up rays of warm sun. His eyes closed shut as he savored a long pull on his cigarette. The nervousness that I had sensed in him in Kodiak seemed to have diminished.

"Well, if it isn't Sleeping Beauty," he said in greeting and smiled.

The *Grant*'s hatch cover had been removed from the raised combing, revealing a big hole about five feet wide by seven feet long. Through the opening I could view the surface of the ice that had been blown into the fishhold in Kodiak.

"Where are we?" I asked.

"We're at the southwest end of Kodiak Island at Alitak Bay. I decided to make a stop here to pick up some more bait from the cannery . . . some salmon tails."

Why, I wondered. Twelve-thousand pounds of bait wasn't enough?

Jack suddenly craned his head up, his eyes directed to a loaded pallet swaying over the deck. It was dangling as though on a thread, hanging perilously from the arm of a dock crane. Vapors of ice-steam cascaded off the block of cardboard boxes. Instinctively, I stepped back, out from under the heavy load.

"Head's up, guys. Here it comes."

I looked down in the fishhold through the open hatch combing and spotted Wally and Chris. They hustled to take cover in the hold, away from the opening, by hiding in the side pens, deep under the deck. Should the crane's cable break or the pallet and cargo fall, they would be safe inside there.

Jack raised one hand, signaling the crane operator, and guided the load by grasping and pulling on the taut cable. The fishhold swallowed the pallet whole as it descended into the open mouth of the hatch combing.

"Whoa," he said, waving up his hand. The load jerked to a halt. "Easy now." The pallet lowered and crushed into the surface of the ice, its weight making the *Grant* sink a little deeper into the water. Jack yelled out, "HOLD IT!"

Wally and Chris flew to work. They jerked the boxes from the pallet and stacked them in the side pens, on top of the bait loaded in Kodiak.

Jack leaned over the hatch. "Hey there," he yelled down into the fish-hold. "Make sure that you stack them in equal piles on each side. I want the boat to sit even-keeled. Don't give the boat a list, okay?"

"No problem, Jack," Wally replied.

In the space of two minutes, the boxes had disappeared from the pallet into the side pens. Jack hollered, "Take it up."

My focus wandered from Wally and Chris loading the bait to Freddy and Kaare working on the stern within the bait tent. Working on piles of rope that rested on elevated platforms, they speared chunks of bait with the points of giant fishhooks, alternating between putting hooks down into the middle of the pile and flopping down a few coils of line. Curious, I walked back to investigate.

"Hey, Slick. Ya have a good sleep?" Kaare asked me with a warm smile. His hands blurred in motion back and forth, not missing a beat while he spoke—grabbing hook, bait, and line—hook, bait, and line—hook, bait, and line. Behind him, Freddy hustled, laying down line and baiting hooks.

"Yeah. What time is it?"

"Three in the afternoon," he laughed.

"How long did I sleep?"

"I'd say fifteen hours or so, at least."

"You're kidding. Whoa. I musta been wiped out." I never slept in late at home.

"Yup. I'll say." He grinned. "Get your skins on, 'cause I'm going to show you how to bait longline gear. You're going to have to learn this sooner or later. Might as well be here at the dock, instead of while we're rollin' around out fishin'."

"Great," I replied, excited, and I spun around on my heels to find my oilskins. I returned, ready, suited up for action.

In the next minutes, Kaare introduced me to the bane and servitude to which longline fishermen are bound—the absolute drudgery of baiting thousands of hooks on very long lines. I'd reached the point of no return.

To me, it had seemed that our mandate was simple: Catch fish by dropping baited fishhooks into the ocean. But this was longlining—emphasize *long*. And arduous. Each day, fresh bait had to be stuck on over 4,000 hooks. Furthermore, this conglomeration of gear needed to be coiled up on the boat, baited, and returned to the ocean without hooks and line tangling. Not simple. The hooks proved to be the sticking point.

"First you need a skatebottom," Kaare began.

A skatebottom is a square of canvas the size of a pizza box, with ropes attached to its four corners—two short ropes about a foot long, and two longer ropes about five feet long. The short ropes have "eyes" or loops at the end; the longer ropes have no eyes.

Kaare showed me how to place the skatebottom on an adjustable wooden platform—the elevator—that hinged down to extend horizontally from the side of the bait table. I took a deep breath and sized up the task. A giant coil of rope sat in front of me, 1,800 feet long—a skate-of-gear, or skate. Gangions attached 70 hooks to the rope, each separated by 26-foot intervals. My job? Bait the hooks. While I went from hook to hook, I reorganized the skate by turning it upside down, baiting the hooks, and making the skate safe to be "set"—put back in the water to fish.

Impossible, I thought.

By the time a few hooks were baited, my skate looked like a humongous bird's nest.

Would this go into the water in a linear fashion? No way in hell. I pictured my skate whirling out past the stern in a frenzy—a brief and furious tornado of rope and flailing hooks. I questioned whether I was cut out for this job.

Kaare looked over to see what I had done. He frowned.

He said calmly, "Okay, let's try this again." He lifted the whole mess and flipped it over, back to where I had started.

I visualized what my skate was supposed to be—a long, linear thread of line dragged out behind the boat to catch fish.

Kaare was reassuring. He said, "This time we'll pay more attention to where the gangions go." Patiently and slowly, he showed me how to layer the gangions between the coils of line for each hook. He said, "Loop the gangions out of the way of the hooks. Less prone to be snagged that way."

In the end, I learned that baiting a skate involved solving a three-dimensional puzzle.

What mattered most was that the skate should go over the stern into the ocean without tangling, strung into a line without snarls.

I worked on my puzzle at a cautious pace, stopping and looking to analyze each piece of bait, studying it for the best place to insert the hook. Years earlier, learning how to fish for trout, I'd been taught to conceal the tip of the hook, and at the same time, to make it lethal. It was a fine balance, and I wondered, dreamily, how a fish would judge my skill. I imagined that I could outsmart a halibut by expertly positioning the hook in the bait. When I had fished for trout back home on the lake, the little trout were about the same size as the chunks of bait I was handling today. My childish fantasies slowed me while I worked through the skate, infusing the job with fun and fancy. The giant fish would be "my fish" because I had baited the hook.

I finished baiting my skate in just over a half hour, confident that "my" skate would catch more fish than any of the other skates baited that day—unquestionably. When I stepped back to admire my achievement, I saw a disheveled lump of rope, loops of gangions sticking out helter-skelter in all directions. Compared with the other skates piled on the stern, my skate was an aberration—an accident waiting to happen.

"Atta boy," said Kaare. "Here, we'll stick this one to the side, so we can keep an eye on it when it goes out the stern."

He pulled on the skatebottom ropes crossing over the top of the skate, reefing and cinching them down much as a cowboy tightens the straps of a saddle on a horse.

"You've got to make these ropes tight, or else you'll have a helluva mess on your hands if the skate becomes loose. You ready for another one?"

I wasn't ready, but I could see from the pile of nonbaited skates that a lot of work remained. I nodded yes and started on a second skate. My feet ached.

Wally and Chris joined us after finishing the loading of the boxes of salmon tails and began baiting skates of gear.

A few minutes into it, Wally turned around to catch me crosseyed, inspecting the point of a fish hook. "What the fuck ya doin'?" Wally growled, interrupting my reverie. "Get a move on."

Jack came about ten minutes later, returning from the fish plant office. I felt some relief knowing the whole crew worked together now, knowing that with the combined effort of six people, I could get away from the bait table that much sooner.

Sometime into my third skate—about an hour later—I became distracted watching a huge boat, more like a barge, come alongside the dock behind the *Grant*. Everyone kept working and left me alone while I took a break at the stern. I sat down on a baited skate to give my feet a rest and watched.

"It's a salmon tender," said Chris, his back to me, baiting hooks without pause. "Salmon fishermen sell their fish to these boats so they don't have to come all the way back to the plant to deliver them."

Not slowing quickly enough, the salmon tender crunched into the dock's pilings with a loud crack. Its crew tied the boat to the dock, hollering and shouting at one another until they had completed the job. They were a disorganized bunch compared to the *Grant*'s crew.

Workers from the fish plant extended a conveyor with metal scoops down into one of the salmon tender's holding tanks. The conveyor began to turn. Scoops disappeared one by one into the tender's hold, dipping down and reappearing laden with big, silvery salmon. At the top of the conveyor, the scoops tipped over. Salmon flopped out to travel along another conveyor belt into one of many warehouses on the shore.

Wally interrupted, chiding me. "Hey now, Flash. It's time to go back to work," he said in a parental tone. "You'll never get any work done with your finger up your nose."

I blushed a little, feeling like my mother had caught me. I stood up and went back to the bait table, to figure out where I had left off.

Damn. Another nickname; yet I liked Flash, Slick, and Kid better than Dino.

As I worked on the next skates, the amount of time I spent fantasizing grew in relation to time spent baiting. On the positive side, I did most of the work myself, except for Kaare retightening the skatebottom ropes after I had tried my best to make them taut. It amazed me how hard he could pull with his arms and back, squashing the body of the skate into a much smaller lump, compact and solid.

For most of two hours, the crew worked in silence. In the third hour, conversation began to percolate, slowly, like a kettle coming to a boil. Talk simmered, gaining momentum, and then the crew came to life. In no time, the conversation roiled crazily out of control, yet the crew's hands never stopped, flying back and forth like darting birds, handling hooks and gear.

The discussion revolved around Freddy and his sexual exploits during the layover. When Freddy's story slowed, Chris prodded for more, wanting him to make it juicier, demanding more detail. Freddy delivered satisfaction.

In comparison, the locker-room prattle back at high school sounded like tales of Dick and Jane. I rolled my eyes, listening intently, pretending to work. Blatantly outrageous, many of his stories focused on hustling women—or, alternately, getting hustled *by* women.

The crew's banter died and rose again several times in the hours that followed. Wally spoke up during an intermission to talk about his family. In just a few steps of conversation, the innocent topic of his boy's baseball game spun raucously into another tale of torrid sex. From baseball . . . to sports . . . to cheerleaders . . . to having sex with cheerleaders—a progression of topics spurred in tag-team style by Freddy and Chris— "You're it."

In six or seven hours, which to me seemed more like twelve, the pile of unbaited gear shrank to nothing. All the skates had been baited. For now, we were done. I helped the crew wash down the working area and tidy up. Jack disappeared down the ladder into the engine room. Kaare and Wally started taking off their gloves and oilskins. I just wanted to sit. My feet felt like bloody stumps.

Jack emerged from the engine room, drenched in sweat, already black up to his elbows in grease and grime. "Hey, listen up," he announced to the crew. "If you want, you can stretch your legs before we leave."

I held back at first, resisting the urge to shout out, "Please, please . . . please, can I escape the boat." Jack read the look on my face, and said, "Okay. Be back in a half hour . . . but, don't be late!" he emphasized, staring me in the eye.

I looked to Wally, who gave me a nod. As fast as I could, I stripped off my raingear. I was still just a kid who lived for this kind of adventure, behaving like I was on a field trip from school.

9

THE
CANNERY
SPY

I STUFFED MY CAMERA INTO MY JACKET POCKET AND SCRAM-bled up the ladder onto the dock. The rest of the crew stayed behind. They gathered alongside the hatch cover, smoking cigarettes, laughing at Freddy's stories. Walking up the dock, I noticed that the pain in my feet had vanished.

I approached some warehouses with a wall of steam rising above them, like images of the Industrial Revolution. A sign next to a doorway said, "DO NOT ENTER." For me, the sign said, "COME ON IN."

I peeked inside the warehouse, a dark space lit only by a series of dust-covered skylights running down the ridgeline of a long, gabled ceiling. I gasped. My eyes caught the glint of what appeared to be thousands, if not millions, of round, golden ingots, stacked waist high. They looked straight out of Fort Knox.

I had found Alitak's "pack"—the cannery's production of canned salmon—the currency of the salmon industry. Cans, and more cans. Lines and columns of shiny cylinders filled the expanse of the room—rows upon stacks—each stack, a mirror image of the other, lined up with another. Something here gnawed at my mind—it was the precision.

I squinted through the viewfinder of my Instamatic camera, attempting to frame the impression. The immensity of the warehouse absorbed the puny flash.

I heard the sounds of people and machinery coming from the next building. I moved on. I followed a forklift that sped in through a doorway, the driver beeping a horn. Forklift traffic was heavy here, as evidenced by a wet trail of tire tracks that led from this doorway into another building down the dock.

Passing behind the forklift, I discovered a busy place showered in the brightness of rows of fluorescent lights overhead. Workers occupied the room amid machinery that clanked and whined. Salmon entered one side of a big whirling machine, and came out the other end dismembered, headless, and without guts. Standing elbow-to-elbow, a line of young women faced a conveyor belt and cleaned salmon. Some trimmed off fins with knives.

I paused to notice that most of the workers were girls only a few years older than me. Faces glowed with rosy cheeks. With their backs to me, they seemed like goddesses. Yellow rubberized aprons gave a hint of their shapes. Triangles of red and blue scarves held back long hair—smelling of fish, I supposed. Other workers, equally demure, scraped and washed away bits of slime and guts, pieces missed by the machinery. Water and fish guts showered off the working platform to flow out through drains in the concrete floor.

My adventure became derailed, simply for the hope of landing a wink of an eye from one of the girls. In a moment, I came back to reality, realizing that fifteen-year-old boys don't earn the time of day from older girls. Recalling that I had a girlfriend at home helped me regain my composure.

Exiting the room, my walk paralleled a conveyor belt with salmon corpses progressing in formation to another room. If salmon can be beautiful, these fish were gorgeous. Their fins and scales were iridescent silver-blue, fish so uniform in size and color that it was impossible to tell one apart from the other.

I strolled through a doorway to the next building and continued on toward noise—a cacophony—in the rhythm of metal clicking against metal. I walked farther. The determined cadence of an industrial factory began to reverberate through my clothing. I heard the sounds of production, pounding and drumming, advancing like a military unit.

Entering the next building, I couldn't help myself and let out a whistle, which didn't matter, because with all the noise here, nobody could have heard it anyway. A maze of intricate machinery filled the width and length of a large room. These were the canning machines—a Rube Goldberg dream come true. The room where they cleaned the salmon had dazzled me with its workers and their femininity. This place rocked with manliness—with machines—lots of machines.

Moving like golden automatons, a line of empty tin cans jerked inside a spiraling metal cage coming down through the ceiling. The cans joined up with a device that wheeled around to grab them. The separate pathways of salmon and can intersected here. Chunks of salmon plopped into each can, which was sent down another path. A woman wearing white factory garb and a hairnet stood next to the canning line and snatched at a few of the cans as they passed by. Using a gloved finger, she stuffed exposed bones and skin deep into the can, so that when a consumer opened the can, the contents would look like solid meat. It didn't bother me that fish bones were in the can. I had eaten enough canned salmon to know that salmon bones dissolved during the canning process. I kind of liked it when I mashed them up with my teeth. My mom told me, "Just chew 'em up. They're good for your calcium."

Now filled with fish, the cans marched off to another mechanical device. An endless line of gold-colored discs—the lids for the cans—descended in a chute from the ceiling overhead. A whirling contraption smashed a lid onto the top of each can, sealing the two together. Another woman in a hair net hovered nearby, motionless except for eyes that flitted back and forth, scanning cans coming out of the machine. I supposed she looked for discrepancies, indicating a bad seal on a can. The cans looked one and the same to me. Indistinguishable. I wondered if this was an Alaskan adventure for her, being stuck in this noisy building.

At the end of the canning line, one of the "Fort Knox" pallets of canned salmon was taking shape. I wondered how many hours of labor had gone into catching the fish, delivering the fish, unloading, cleaning, processing, and canning.

But wait a minute. The salmon inside the can wasn't cooked. It was raw. I knew that the salmon had to be cooked.

I looked around the room. The workers were fixed in place. The progression of the canning process stopped at the pallet. Where did the path lead to next?

I searched for some sign and found forklift tracks next to the pallet of cans. Ground deep into the planks of the floor was a crescent-shaped indentation. Over the years, the tires of forklift trucks had worn a depression where they backed up to one side, turned, and pulled forward to pass through another large doorway. I followed the arc of the track and ambled into the next building.

At first, I couldn't figure out what I was seeing—six giant steel cylinders, five feet in diameter and about twenty feet long, lying side by side, recessed about a foot or two into the concrete floor. The door on one cylinder was open. Inside, four pallets of cans almost filled the huge tube, leaving space for one more.

I walked up to a man wearing a hard hat, who peered through grimy spectacles at a clipboard. He squinted and read one of the gauges affixed to a steaming cylinder. In my face I felt the heat of steam hissing out of a nozzle.

To catch the attention of the man, I simply pointed at the cylinders.

Seemingly unaffected by my intrusion, he yelled, "It's a 'retort,' another word for 'big pressure-cooker.'"

A gargantuan cooker, I thought to myself.

"Darn big," he added. "In just ninety minutes, we can cook 4,000 pounds of salmon in one of these things." He patted the cylinder like he patted the head of a child. Lifting his arm, he checked his watch and wrote the time on his clipboard.

Time! . . . dammit!

Urgently, I asked the man, "Can I see the time?" He raised his wristwatch up to my face.

I had put myself at risk of missing the boat.

I said, "Thanks!" and spun to the door, running in a full sprint.

By the time I neared the end of the dock, my lungs sucked for air. At seeing the *Grant*'s masts poking above the level of the wharf, a wave of relief washed over me. I leaned over the dock to peer down at the *Grant*. I saw heads turn to look up.

My uncle yelled, "We were beginning to wonder if you had gone AWOL on us," adding a little chuckle. "Wouldn't be the first time you know."

His face turned grim. "Now get your ass down here, goddammit!" he ordered. I complied and descended the ladder in a hurry. My feet landed on the deck.

"Okay, guys." Jack cried out. "Let's get goin'. Throw off the lines." Impatient, he hollered again, "Let 'em go."

I took the position amidships, to handle the springline. Freddy manned the bowline, and Kaare took the stern. I freed the hitch holding the knot tight and flipped figure-eights off the cleat. I made the line slack, got the looped end free, and let it drop into the water. I pulled hard on the line, coiling it into a heap at my feet. The rope snaked through the water around the piling and slid into the boat.

Jack gave the engine some throttle and jockeyed the boat back and forth a couple of times to bring the stern away from the dock. In reverse gear, the *Grant* pulled away backwards, becoming clear of the dock. Jack put the boat back into forward gear and turned the wheel, and the bow swung round to head out of the bay into clear water.

The *Grant* heeled over slightly, swinging around Cape Alitak—the last headland of the bay. Until I got my sea legs, my future would be uncertain starting now. That is, *if* I got my sea legs. Solid earth would be absent under my feet for the next three weeks. Jack stood vigilant, steering from the pilothouse.

Our last vestige of land, the Trinity Islands, lay off to our left, or port bow. Filtered through an atmosphere of ocean haze, this group of low-lying islands appeared as a pastel-purple smudge pressed flat onto the azure horizon, colors softened by the orange of the setting sun. Another wave of excitement hit me as I felt the swell of the Pacific again. Nothing but open water stretched in front of the boat.

I noticed that I stood on deck by myself. The rest of the crew had disappeared. I crawled down the ladder and went to bed. The distant rumble of the engine and the vibration of the propeller and shaft eased me to sleep.

We traveled through the night into the next day.

10

SETTING
GEAR

SOMEONE SHOOK MY LEG HARD, WRENCHING ME OUT OF A
deep sleep. Never before had I been roused this way. At home, my
mother gently nudged me out of slumber if need be. On a normal day, I
woke up on my own.

Waking, my mind went in slow motion. I recalled hearing a voice, but
the words were jumbled. With one eye open, I peered out of my bunk to
investigate. I saw my uncle shaking Chris and Wally, making their bodies
roll when he shook them. Duty called.

The boat lurched from side to side. Dishes clinked. The stove purred.
Lying down, I heard the sound of water right next to my head, slapping
against the planks of the boat. I gathered from the way the boat rocked
that we were rolling in the trough—in the valley of the waves. A swell
heaved the whole boat up and down—up and down—up and down. The
fo'c'sle moved like a carnival ride. Under these conditions, getting out of
bed concerned me. No ladder reached up to my bunk. I'd have to climb
down.

To start, I found a handhold and a notch to place the edge of one foot.
Pushing myself out of bed, I moved with my arms and legs splayed out.
A contortionist would do well here, I thought. I was still a couple feet

from the deck, when the boat seemed to fall into a hole in the ocean. My foot slipped. I swung forward, hanging from my hands as the bow dove down. My head bounced off the corner of a deck beam, making a thunk like a melon. I cried out in pain as my hand shot up to nurse the lump emerging from my temple. This left me with just one handhold on the boat, which made me swing out of control. I was forced to quit massaging my head to complete my descent.

"Nice job," Wally chuckled, from up in the peak. He put on the last layers of his clothing, sitting safely within the V where both benches of the galley table came together in the bow. I forced a smile.

"Make sure that you put on some warm clothes. We'll be on deck for at least a couple of hours," he said.

Grateful to have landed on the galley bench, I wedged myself between the table and the bunk behind me. Hunched over, I rubbed my head.

"Dean. Look out."

I looked up and flinched to find Chris towering over me, standing on the galley seat, trying to make his way toward the stove. Sitting here, I blocked one of the two available paths out of the peak, one down each side of the table.

"Just lean forward."

I bent over, driving my nose into the rubberized pad that covered the galley table. He stepped over my back to the end of the bench and leapt gracefully to the deck, timing his jump to coincide with the boat falling off a wave.

Getting dressed was no small affair. I put on two sets of underwear—an inner and outer layer—a turtleneck shirt and wool clothing, both pants and a jacket. Layer upon layer.

My fellow crewmen wore clothing of thick wool that covered them from head to toe. When I prepared for this trip, I had packed a pair of thin wool slacks that my mom found at a secondhand store. My mom shied away from coarse wool in favor of the finer material of the worn-out slacks, believing that they would suffice during the summertime. They would also be more comfortable by not being itchy. My workpants had the density of a worn-out handkerchief.

Everyone on the *Grant* protected their hands with two pairs of work gloves. The inner pair was a common, yellow, rubber dishwashing glove. The outer cotton glove, "green gloves" as the crew called them for their color, protected their rubber gloves from punctures and their hands from cuts. I had a difficult time getting the work gloves on my hands, wrestling with both pairs.

For footwear, Jack had lent me a pair of his three-quarter-length rubber boots. Long and stubborn, the tops of the boots ended at my crotch. They felt like splints on my legs. Appearing for my first shift of sea duty, I climbed up the ladder into the dim light of early morning, walking across the deck as jerky and stiff-legged as Frankenstein.

A heavy mist drenched the boat in a soggy veil, reducing visibility to less than 100 yards. It was a dramatic change from the sunny weather and the expansive view of the ocean when we had left Alitak the day before.

All things glistened, varnished with moisture that painted a sheen on dull surfaces. A string of light bulbs lit the deck and ran laterally along the boom, above the deck, and between the masts, enhancing the red-orange of buoys and flags—vibrant colors compared to the gray ocean surrounding the boat. This was strange, I thought. Back at home, fog and drizzle didn't occur in the middle of summer.

I trudged toward the stern, up the steps to find my oilskins, hung up inside the bait tent. This final layer—pants and jacket—presented me with another ordeal in getting dressed. To start, I lifted my boot into the plastic pant leg and nearly fell down when the boat rolled. The flimsy pant legs clung to my boots like plastic wrap. I crammed my butt into a corner and tried to get them on. I tugged and pulled, finally forcing my boots through the legs. Compared to what I had worn at home a few days ago—a pair of shorts and a T-shirt—this felt like a suit of body armor, or a straitjacket. Fully suited up, I struggled to move about the deck. Within minutes, I needed to sit down again and take a rest.

Jack barked out instructions to the crew, leaning his head out one of the tiny pilothouse windows. "One-hundred fathoms . . . no lights . . . ten skates per string. Dean . . . you stay out of the way."

The only part of what he said that made sense to me was "Stay out of the way."

The crew went into action. Kaare and Freddy passed by me on their way to the stern. Kaare stopped, noticing the vacant look on my face.

"Hey, kiddo. If I were you, I'd sit inside the companionway and watch for a while."

"What's goin' on?"

"Well, first, we're going to set the gear, which means put it into the water—probably six strings of ten skates each."

"When do we catch the fish?"

"The gear sinks to the ocean floor and we let it soak for a few hours. The bait attracts the fish, and the fish bite the hooks. After a few hours, we start hauling the gear back—one string at a time. As we retrieve the gear, we bait it again, so that it's ready for the next day."

"But how do we haul the gear?" I asked. "Aren't the skates attached to the stern when we fish?"

"To the stern?" he replied. "I don't understand what you're saying."

During my childhood, I believed I knew much about what my mother's family did for a living—fishing with longline gear for halibut. As it turned out, I had many misconceptions of longlining. I had become confused as a child when my uncles' boats were in Seattle, unloading halibut. I had only seen stowed longline gear, bound up and piled in a heap over the aft deck. At the time, I thought I had longlining figured out. I was wrong. I asked more questions.

"You know, Kaare . . . like trolling for salmon, or for trout. Don't we troll the skates behind the boat?"

In the years before, when I had looked at the gear stored on my uncles' boats, all I could see were hooks and rope. I supposed that the ropes were set—run out from the stern—uncoiled to be a single line towed from behind the boat, trolled in the same way that my brother and I trolled for trout on the Lake back home.

Kaare couldn't help himself. He bent over, laughing, "No . . . no. . . . It's *not* attached to the boat after we set it. We let go of the gear. We anchor it to the bottom and mark it with buoys and flags. You understand, kid."

Let the gear go? This was crazy. . . . Let it go free in the ocean?

I fudged, "Thanks, Kaare. I think I've got it."

Kaare left me and made his way to the stern.

I shook my head as I tried to reform my understanding of longlining. It seemed reckless to let go of the gear. It could be lost that way.

Wally and Chris climbed to the bow to untie coils of rope lashed to a pipe railing on the port-side gunwale. From the starboard side, they untied a couple of red buoys—called bags or buoybags—three feet in diameter. On top of each bag was painted a big letter G—for *Grant*. Different numbers were painted on each bag, 4 . . . 12 . . . 9 . . . in broad strokes of deep indigo blue.

On top of the wheelhouse, the radar turned round and round, shooting an invisible beam through the wall of fog. The sound and vibration of the engine slowed and the boat started moving forward, bobbing gently through the sea swells. Jack had put the boat into forward gear. Back on the stern, Kaare and Freddy bent over some skates.

"Okay, guys," Jack hollered down to Wally and Chris on deck. "How are you doin'?"

Wally cupped his gloved hands and yelled back toward the stern, "Fred. You ready?"

"Yeah. Go ahead."

"We're ready, Jack."

"Okay, hang on. I've got to turn the boat."

From on deck, I watched Jack's head in the window frame, bobbing in and out of view as he spun the wooden wheel, four feet in diameter, round and round.

Made of teak, the wheel was the most ornate object on the *Grant*. I never heard anyone refer to it as a steering wheel. Its spokes had been tooled on a carpenter's lathe, shaped in a colonial-style design. Finished with thick coats of varnish, the spokes radiated out from a brass hub that had long ago turned green. Corrosion from the salt-air environment had changed its shiny metal into a rough patina—verdigris. Many times as a kid, I had played with the wheel and knew that it took seven revolutions of the wheel to move the rudder, submerged deep below the *Grant*'s stern, from one hard-over position to the other. I had counted them—right to left—left to right. Turning the wheel, from one extreme to the other, had exhausted me.

Jack's head bounced up and down again. The boat's path straightened. He hollered, "Let her go," and gunned the throttle of the engine. From the top of the pilothouse, the smokestack gave a puff of gray smoke. The boat gained speed through the water.

Wally grabbed a bamboo flagpole. Ten feet long, it had a flag made of orange canvas on top, and on the bottom, a sash weight of about eight pounds of iron. The flagpole had four floats along its middle, each the size and shape of a football. Wally raised the flagpole up like a jousting lance and threw it overboard. After splashing into the water, the flagpole quickly bobbed upright, kept afloat by the floats. The flagpole was tethered by a rope to a buoybag, which Chris held in his hand. The *Grant* moved faster through the water now. Just as the tether became taut, Chris heaved the bag overboard.

Wally dodged out of Chris's way and hustled over to the hatch cover in the center of the deck to grab two shots of buoyline attached to the bag in the water. He heaved the buoyline onto the water's surface, one shot at a time. I knew that each shot of buoyline was fifty fathoms long. Wally had tossed 600 feet of rope into the ocean. In coordinated movements, Wally backed away from the rail and Chris moved in, leaning down to grab the shaft of a fishing anchor.

A halibut-fishing anchor is a typical anchor, with a T-shaped stock on top and two curved arms making a U on the bottom. The flukes at each end of the arms, looking like tips of broad steel arrows, were designed to dig into the ocean bottom. From my nautical handbook, I had learned that this kind was called a kedge anchor.

Taking most of a minute, the boat reached top speed—eight and a half knots. The *Grant* plodded like an elephant, her engine trumpeting pachyderm power and speed.

The other end of the buoyline that Wally had thrown overboard was tied to the anchor Chris held. He lifted the fishing anchor over the side of the boat until it rested by hanging on the rail. He waited, holding the anchor so that its flukes hung a couple of feet above the surface of the water that raced alongside the boat. This looked dangerous to me. The anchor seemed like it could slip off the rail at any moment and fall into the water.

The buoyline came tight and jerked the anchor in Chris's grasp. Casually, Chris lifted the anchor and heaved it overboard, yelling, "ANCHOR OVER!"

In a few seconds, I heard the clink of fishhooks flying out the back of the boat, clicking against the metal of the setting chute on the stern. It took me a moment to realize that our fishing gear was being dragged into the water. How was that, I wondered. Had I missed something in the tangle of ropes? How did the anchor connect to the gear on the stern?

"Is it okay for me to go back there, Jack?"

"Yeah, just stay out of their way."

About halfway through the bait tent, getting closer to Freddy and Kaare on the stern, I heard a loud, high-pitched noise that made my teeth ache. The sound stopped, but I flinched at the frightening spectacle of setting gear. Fishhooks tore through the air, one after the other, flipping out the back of the boat. I heard the sound again and saw the tip of a fishhook make a spark as it scraped the sheet metal of the setting chute, screeching like chalk on a chalkboard.

The physics required to set longline gear out the stern of the *Grant* were simple—the boat moved and, consequently, the gear was dragged out the back of the boat through a chute designed to allow hooks and line to slide over the stern rail without getting snagged. At the base of the chute was a five-foot-long horizontal ramp with rails a foot high on each side to brace the skates and keep them from tipping over as the boat rolled.

Playing and climbing on the *Grant* as a kid, I puzzled over the function of the chute. It looked harmless—an elegant sculpture of a wave, about to break over the stern to fall into the water. No longer benign, the chute now horrified me.

Freddy stood forward of the chute, leaning toward it. He looked dangerously close to the hooks flying out. He was bent over with his back to me and used his hand to push a skate directly into the chute. It was easy for me to imagine a hook catching him and dragging him over the side into the ocean.

My mind flashed from one horror to the next. If the hook snagged Freddy's arm, would the hook tear out of his flesh? Or, would he be pulled overboard and sink to the bottom with the gear? If Freddy screamed, could Jack hear him over the deafening roar of the engine? If Jack did hear him, how quickly would the boat come to a halt? I knew how long it took for the boat to speed up. How long would it take for the boat to come to a standstill? Could the crew and the boat possibly react in time?

The gangions added to my anxiety. The three-foot-long gangion made each baited hook whip out of the chute like a nonstop mechanical bullwhip, one after the other. The baited hooks were flung into the sky, then they swung down and smacked into the water with a splash. I winced at hearing the chalkboard sound again.

Behind the boat, sea gulls sailed like kites on strings, dodging back and forth to gobble up scraps of bait that sprayed into the air and onto the water.

Kaare stood behind Freddy and supplied him with fresh skates, the ones that we had piled around the stern after baiting them in Alitak. Kaare slung the sixty-pound skates around the stern with ease, as though they weighed only half that much. After hefting a new skate into the chute, Freddy untied the ropes lashing the skate bottom. He removed the skate bottom by yanking it out hard and quick from under the skate, like someone pulling a tablecloth out from under dishes on a set table. He did so to avoid disturbing the skate and the baited fishhooks arranged with care inside. He tied the next skate to the bottom line of the skate going out the chute. Freddy and Kaare braced themselves to stand as the *Grant* reacted to ocean waves. Every two minutes, a skate disappeared out the chute.

At times, the gear jumped up inside the chute, launching from its coils like an attacking snake, sometimes right at Freddy's head as he leaned down to push in a new skate. He jerked back, and so did I, regardless of my safe distance away from the gear. Then, it hit me. As I learned more skills as a fisherman, I would be expected to set gear. I would stand in Freddy's place, doing exactly what he was doing. Could I do that job now? No way in hell.

I watched skate after skate shrink to nothing and get replaced by a new skate. I sat down to rest and to watch. The twisting line going out the chute looked like a twirling lasso. I stared, transfixed.

Twenty minutes later, Freddy turned toward the bow and shouted out over the roar of the engine, "LAST SKATE IN THE CHUTE!"

Ten skates had been set—the first string.

While I was away, Chris and Wally had busied themselves on the forward deck. Coils of line covered the hatch and stretched over the deck to anchors and bags and flagpoles. Wally grabbed a huge coil of rope that had one end leading to an anchor, lifted the anchor, and hung it over the rail, as Chris had done. I could picture the arrangement of the gear in the ocean, but how was all of this line going to make it overboard in an organized fashion?

"ONE LINE!" Freddy yelled from the stern.

Wally repeated the order, yelling "ONE LINE!" to Jack in the pilothouse.

Jack pulled back on the throttle. The engine's roar subsided, and the *Grant* slowed—all had become quiet. I noticed that my ears were ringing.

From the stern, Freddy yelled, "END OVER!"

Looking through the bait tent, I saw Freddy flip a line over the top of the chute. Just as he did so, Wally heaved the anchor into the water with his right arm and threw the coil of line over the rail and into the water with his left. Chris stood behind him with the bag and flag. Facing aft, he threw the buoybag into the water with a sweep of his right hand. Then, with both hands, he threw the flag into the water, right behind it. Very impressive. One half of the ropes had disappeared over the side in less than four seconds. The coordination between the bow and the stern and between Chris and Wally amazed me.

"Hey, Jack," Chris yelled. "Are we setting end-for-end"?

"Yeah, go ahead," he replied.

Chris turned and grabbed another bag and flag, tossing them into the water to start the second string. The engine of the *Grant* spun up to reach maximum revolutions. A bow wave slowly rose up and spread to curl around the boat, growing steep, then collapsing along its flanks with a hiss. Chris threw the last of the maze of ropes into the water. Wally lifted the next anchor up to the rail.

Wally yelled out, "ANCHOR OVER!" I heard him grunt as he flung the anchor into the water.

For the time, I chose to stay forward, on the main deck. I was eager to distance myself from the commotion on the stern and ready for a break from watching Kaare and Fred. Then, seeing Jack move within the tiny pilothouse, I decided to pay him a visit. He leaned his head out an open window.

Stepping up, I asked, "Where are we?"

"What's the matter? You lost?" he answered, grinning.

"Nah, where are we?"

"Young man," his tone was one of teasing bravado, "You . . . are . . . in . . . the middle of nowhere."

"Come on, Jack."

He was clearly excited to be fishing again.

"Well, you *are* in the middle of nowhere—off the western Alaska Peninsula, a little ways past Foggy Cape."

How appropriate, I thought, looking at the fog.

"Other points of interest around here are Castle Cape and Chignik Bay."

Here? What was he talking about? A landlubber, I still related to geography as solid things in terms of land. "Here" describes a tangible place, a location with dirt, rock, or soil. There was nothing *here*, just fog and water. I needed a position, a latitude and longitude.

I looked across the water, what little of it I could see. Though the ocean's surface moved a lot, its surface was smooth—with no ripples from wind. The fog had thinned a little, which only further confused things by blurring the line between fog and ocean. I couldn't tell where the sky stopped and the sea began. Without any visual references, I felt dizzy. Vertigo. I needed to restore my equilibrium by turning back to something I could focus on, to the pilothouse and to Jack.

I asked him, "How far away is the horizon . . . when you can see it?"

"The horizon? The distance of the horizon depends on the height of where you are standing. Boats the size of the *Grant* will drop below the horizon when they are only about eight miles away. If you're lucky, you can see some masts sticking up over the horizon at distances greater than eight miles. So there you go. . . . I guess that would place the horizon out halfway—just four miles away."

I floated on board the *Grant* with the ocean sloping out around me in all directions, down past the shrouded, invisible horizon. I imagined I was on top of the world. With that thought, I smiled, feeling better.

For a moment Jack withdrew from the window, but he reappeared, looking harried.

"Listen up, old buddy, ol' pal. I've got to figure out what I'm going to do with the next string." His face spread into a grin. "Buzz off, kid." He ducked into his stateroom to pour over his fishing chart—a nautical chart with a bunch of pencil marks on it.

I didn't buzz off but instead stuck my head in through the vacated window. Jack was scanning the chart and instruments, looking up to the *Grant*'s compass and over to his fathometer—also called a depth-finder, depth-sounder, or simply sounder for short. The readout on the fathometer screen looked like a recording of an electrocardiogram. The jagged series of spikes and valleys represented the ocean bottom. The challenges that Jack faced to find and catch fish intrigued me. I imagined being under water and seeing what the ocean bottom looked like—out here.

My uncles had shown me how the sounder worked when I visited their boats in Seattle. Looking from the window, I saw the faint electric spark that flashed each time the rotating armature came down, burning a small black mark on the paper. At a snail's pace, the fathometer paper gradually spooled out through the glass-covered display, and at the same time, rolled up onto a spindle on the other side of the screen. The series of marks created a scribbled image across the paper, giving an impression of the seafloor passing deep below the *Grant*.

Jack emerged from his stateroom.

"Oh, you again," he said, reaching for his smokes.

"Is this a good spot, Jack?"

"That's a good question," he replied.

I knew Jack as a champion of verbal jabs—a Mohammed Ali of quips and wisecracks. I braced myself for another poke. None came. He answered me with sincerity.

"With halibut, you just never know. What I *do* know . . . is that I have caught many fish here in the past. We just won't know how good it is until we haul the gear back later today." He paused to draw on his cigarette. "Halibut are tough to find—they're invisible," he said, looking over to the sounder. "Some people say fathometers are fish finders. Unfortunately, fathometers are not halibut finders. Halibut don't have an air bladder. It's the air bladder that shows up on a fathometer."

Jack leveled his gaze. "Hey, punk. Don't you have anything better to do? You've got to spend some time on deck before you can pretend you're the captain. Get out of here."

I dropped out of the pilothouse window to watch Chris and Wally get the buoyline of the next ends ready.

The scene on the forward deck felt much safer than the turmoil on the stern—the engine noise, the scraping of hooks, and the threat of getting hooked. Seafoam hissed along the sides of the boat, which rolled slowly and gently from side to side. Gulls cavorted above, taking turns to break out of formation and smack into the water to snatch a morsel of herring flung off a hook. The air was thick with the fishy smell of bait.

"Are you sick yet?"

I whipped my head around, "Wha'?"

It was Jack again, leaning out the pilothouse window, staring vacantly at the ocean. A cigarette stuck out the side of his mouth.

"No, I'm not." His question irritated me.

"Good . . . that's good," and his head disappeared back into the pilothouse.

My gut tightened just thinking about it. I hated throwing up—always had, hated it more than anyone in my family. I'd do anything to avoid vomiting. No matter how sick I felt, I shunned my mother's advice: "Come on, honey. You'll feel better if you do it."

I spent most of the next hour and a half alternating between the main deck and the bow, watching Chris and Wally set up the buoyline. The largest uninterrupted space on board the *Grant*—the port side of the main deck—was six feet wide by twenty feet long, large enough for Chris and Wally to organize and set a mile and a half of buoyline along with a dozen bags and flags.

The bow deck was small and felt scary. Its railing was only as high as my knees, so equipment on the bow provided many handholds—mast cables, the ventilation funnel for the fo'c'sle, pipe railings, an assortment of ropes and lines. The smokestack of the galley stove felt warm under my two layers of gloves. Compared to the stern and the main deck, the bow reacted to the sea swell with exaggerated movement, alternately soaring about ten feet into the air, then sinking into the waves, making it seem like a roller coaster ride—but without a seat or lap bar.

Wally swung from a cable down to the deck, looking like a kid playing on a jungle gym. Chris did the same when he came down. Lots of agility and movement.

I'd like this part of my new job. With something to hang onto up there, it struck me as great fun. I loved to climb trees. No matter how far off the ground, I always felt safe if I had my hand clamped to a branch.

"How's it goin', kid. You feeling all right?" Wally asked, turning toward me.

"Yup."

"That's good," he said, and went back to his work.

Nervous and self-conscious, I felt my stomach twist in my gut. I wished they'd quit asking me how I felt. To myself I pleaded, please don't get sick.

Almost two hours had passed since we started setting gear. I guessed that we would finish soon when I noticed only one coil of buoyline lying on the hatch cover.

For the sixth time that day, Freddy hollered, "ONE LINE!"

The engine slowed. Chris threw the last bag and flag overboard.

The last end of six strings had disappeared behind the boat into the fog. In two hours, Wally and Chris had tossed over the rail twelve fishing anchors, twelve bags and flags, and 7,200 feet of buoyline. Stationed on the stern, Kaare and Fred had sent 108,000 feet of fishing gear out the chute—twenty miles in all—along with more than 4,000 baited hooks.

The main deck looked spacious and tidy without the anchors and buoyline. On the stern, instead of being packed from rail to rail with skates piled high, the aft deck looked like half of a small, round dance floor. In my life, I'd never seen this area emptied of skates.

Kaare washed down the stern with a deck hose. Freddy shoveled several pounds of bait out of the chute, pieces having come loose during the process of being set. Up forward, Chris disappeared through the small hatch into the fishhold and started hurling boxes of frozen bait up to Wally, waiting above. In a couple of minutes, they were finished and began to take off their gloves and oilskins.

I felt smug. Being a fisherman was no big deal—I could do this work. I understood the different roles of the crew in setting gear. Other than dealing with serious issues around setting gear on the stern, I figured that I could do these jobs—no problem. Chris and Wally's job, setting up the buoyline on the forward deck, looked to be downright fun.

The engine sped up and we traveled at full speed. Jack flew out the pilothouse door and landed on the deck with a thud. "Okay, guys. I'm going to run the gear down, back to the first end. We'll be getting started in four hours."

"Dean . . . you get to bed right away. You've got a long day ahead of you."

"Okay, Jack."

A mutual grunt echoed back from the crew.

I looked forward to the next phase. I wondered if the first string set in the water today had caught any fish yet. Romantic images danced in my head: I pictured myself and the crew congratulating one another, celebrating the successful capture of yet another giant halibut. This setting gear stuff was great, but the next step, hauling gear and catching fish, must really be super.

I crawled into my bunk and started a letter to my family. I wrote in a notepad I had made as part of an art class in school. I had decorated the paper with a silk-screen image of Schultz's Snoopy.

7th of July—Still not sick! Woke up yesterday at 3:00 PM, saw a couple of porpoises (white belly) and a couple sea-lions. Stopped at Alitak Cannery. Learned how to bait, went to bed early. Nice weather.

Little did I know.

PART TWO
First Day of Fishing

11

AWAKENING

I LAY WIDE AWAKE IN MY BUNK FOR WHAT SEEMED A COUPLE of hours, surveying cracks in the paint of the deck beams and planking two feet above my nose. Over the previous five decades, the texture of wood grain had been lost under layer upon layer of enamel paint—initially white, now forever stained with soot and smoke. I wondered who lived in this bunk before me.

Not being able to sleep made me more anxious, which added to my sleeplessness—a catch-22. The same rush of excitement had fueled my insomnia before opening day of trout season on the Lake at home—the thrill of the chase. I tried to rest, feeling safe within my little bunk, my cocoon.

I heard the engine slow. I relaxed as Jack took the boat out of gear. Noises of water and boat subsided, and we began to drift, bobbing on the open ocean. I was learning about the sounds, motions, and vibrations, as well as the quiet times—what they meant. From my bunk, I could already tell if the boat was in gear or not, feel the size of the waves, and even determine which direction the waves traveled in relation to the heading of the boat. In the end, the rocking of the *Grant* won, more powerful than insomnia. I fell asleep, lolling and rolling in the sea swell.

It seemed only a moment later when I received the same annoying jerk on my leg that had roused me earlier in the day. However this time, my awakening was accompanied by something new—Jack, reciting a rhyme in a coarse tone of voice.

"Drop your cocks
and grab your socks . . .
It's halibut hauling time."

"Good god," I muttered. I inched out of my bunk and began a slow and calculated climb down to the deck.

The crew moved swiftly and spoke few words, as though they had something on their minds. I sensed a state of heightened energy. By the time my feet reached the floor, Freddy stood at his station next to the diesel stove, which roared. Whining loudly, the fan blew air into the firebox, sounding like the combustion chamber of a turbojet. Freddy's dense mop of black hair was slicked back, greasy with sweat.

Now that we were offshore, the boat rolled nonstop. I saw that metal railings had been attached to the perimeter of the stovetop. Brackets attached to the railings held a large steaming pot, trapping it on the stove. Similar wooden rails partitioned the galley table as well. Freddy inserted a metal handle into a hole on a circular, pancake-size iron disc built into the stovetop and lifted it off the stove. As he removed the disc, a reflection of the inferno inside the firebox danced across the stainless steel wall behind the stove.

Freddy reached for an odd-looking pot hanging from a hook on the wall. Instead of having a flat bottom like a normal pot, it had an extension out the bottom, six inches long, narrower than the rest of the pot, blackened and covered with soot. He filled the pot partway with water and carefully set it on the stove, with the protrusion fitting into the hole, extending down into the inferno.

Curious, I asked, "What's that?"

"It's the hotpot," he answered, keeping his focus on the stove.

"Oh."

Freddy was in no mood to chat.

He spun around and grabbed a large tin of ground coffee. With a large soup spoon, he shoveled heaping scoops of coffee out of the can and into the hotpot. I decided to quit asking questions and finish getting dressed.

Before I had put on the rest of my clothes, the hotpot rolled in a froth. The rich vapor of boiling coffee smelled nice, cleansing the air of odors— of humans, food, and fish.

Moving slowly with concentration on his face, Freddy lifted the boiling brew off the stove and poured the coffee through a sieve into a metal coffeepot held on the stove by railings. He gave a few quick strokes on the handle of the water pump to rinse out the hotpot and returned it to its hook on the wall. Freddy reached up to a row of stout mugs that hung from brass hooks stuck between the deck beams, just inches above his head. The cups swung back and forth with the roll of the boat. He filled the cups, one after the other, and sent the brew to the crew, sitting in silence along the pie-shaped table.

"Hey, Freddy. Grab my smokes, would you?" Chris piped up.

"Yeah, mine too," said Wally.

Kaare just nodded.

Freddy climbed up the fo'c'sle ladder a couple steps and reached backwards to a small wooden shelf mounted high up within the companionway. Three packs of cigarettes flew through the air, bouncing onto the tabletop. Stick matches cracked into flame. Within seconds, a cloud of cigarette smoke filled the entire space of the fo'c'sle, a haze so thick that halos appeared around the naked light bulbs that illuminated the galley. Looking as though they were in some kind of trance, the crew stared at the steam rising from the coffee cups in their hands. The smoke thickened until the scene became a surreal blur.

Freddy broke the silence. "You want a cup of coffee, Dean?"

Sitting at the most forward end of the table, farthest from the stove, I replied, "Sure, I'll try some."

"Try some? You mean you haven't drank coffee?"

"Well, maybe a couple times."

Freddy poured. "Jeezus. We've got ourselves a real greenhorn here." He smirked to the others around the table. "I suppose you're a virgin as well."

Oh god, I moaned. I blushed even harder, anticipating this question from these guys at some point.

"Well?" he prodded.

"Maybe."

A couple sniggers.

"Maybe? My ass. Come on, are you a virgin or not?"

"What if I am?"

"Well, if you are, we got to do something about it."

Chris stirred, fondling his cigarette, "Damn right." He smiled.

"What are you talking about?"

"What I'm talking about is that there is some really sweet stuff in Kodiak."

"We could get you one," Wally piled in.

"Get me one what?" I was starting to sweat now, and not from the heat.

"A hooker, for crying out loud. Chrissakes, you really did just fall off the turnip truck, didn't you?"

"Shit. You'll probably make enough money from this trip that you can afford one . . . or better yet, we can all chip in. How 'bout that, guys? That'd be the right thing to do? Now wouldn't it?"

"Yeah."

"Sure."

"You betcha."

"Oh, yeah."

Finishing the last bites of a bowl of cereal, Chris had milk dripping down both sides of his mustache into his beard. His heavy glasses obscured his expression as a rule, making his eyes dim behind thick lenses, but now they glinted like sparks. "But we'd get to pick . . . wouldn't we, Freddy? What do you think, the black one?"

Someone snorted. Another slapped his knee. They began to roar, shaking from laughter. A caterpillar-size ash fell from Wally's cigarette onto the table. By their antics, I began to sense this was all just a big prank and relaxed a bit.

I tried to improvise a laugh, but stopped halfway through a counterfeit chuckle. I cringed, looking over to Freddy. His eyes had narrowed and a sweet smile spread across his face, a look that erased any doubt in my mind. There really was an African American hooker in Kodiak. They were going to chip in. This wasn't a joke.

The discussion escalated into a full-on debate as to the ultimate way to snuff out my virginity. They had a great time of it, elbows on the table, their heads and shoulders heaving, taking turns at adding another bit of detail to the fantasy they created for me.

"Goddamn best thing for you."

"I wish I had a hooker for my first."

"We'll make sure she takes it easy on you."

"Your girlfriend will love you for the things she'll teach you."

"Makes me weak just thinking about it."

With a smack, the sole of my uncle's boot hit the top step of the ladder. The conversation halted. Jack descended into the fo'c'sle and glanced around the table. "How 'bout a cup?" he asked, reaching out with his mug toward Freddy, standing by the stove. By the way he looked at the crew, I think he sensed something. Evidence lingered in the fo'c'sle, painted on reddened faces. My face tingled from blushing so hard.

Jack chose to be oblivious. He took a deep breath, and with an extra measure of enthusiasm in his voice, he announced, "Time to get goin' boys. We're only a couple of minutes away from the gear." Without delay, the crew snuffed out cigarettes in the empty butter-cans that served as ashtrays. Jack disappeared up the ladder with a steaming cup of coffee.

Still sitting, they slid down the galley seats one at a time, stopping upon reaching a small space at the base of the fo'c'sle ladder. The others paused, waiting for their turn to stand up, to hang up their coffee cups on hooks, and go up on deck.

I occupied a seat at the peak, so I was last to get up and get ready. A fine tactical position, I thought, with my logic askew. After all, I didn't know what to do. Right?

As it was, my arrival lagged five minutes behind the rest of the crew, having frustrated myself, fumbling with getting on my clothes, boots, and work gloves—those damned gloves. I exhausted my hands even before I began to work, struggling to put on my gloves.

I appeared on deck just as Jack gunned the throttle of the engine, backing down in reverse gear. The boat rumbled and shook. Clouds of bubbles swirled in whirlpools that wrapped around both sides of the stern. The propeller tore into the ocean, struggling to slow the massive boat.

I had been waiting for this. We would catch a fish any moment now, I thought.

"Hey, Fred," yelled Wally. "Bring me a smoke when you come up on deck."

"Me, too," added Chris.

Wally stood with his knees braced against the starboard rail, his body arching out over the water. He held a long bamboo pole with a large, menacing hook on the end. To his right, the entire starboard side of the deck had been fenced with boards into four sections—two large and two small—in a configuration for fishing. Like this, the deck seemed cramped for space. The hatch combing and cover, a wooden block the size of a large dining-room table, sat in the middle of the deck and doubled its function to become a work table.

Jack took the boat out of gear. Silence fell across the deck. The boat dipped slightly into a small sea swell, and I caught a glimpse of something orange traveling toward us, just over the bow railing. Wally swung the pole forward over the water. The tip of a flagpole appeared and slid toward Wally. The flag slapped into his shoulder, nearly hitting him in the face. Once more, the engine roared, bringing the boat to a complete halt.

Kaare took the long hook from Wally, tossed it up to the bow deck, and shoved the flag and bag over to one side of the deck. Both Kaare and Wally turned to pull the gear aboard in unison, hand over hand, through the roller. The roller is a heavy metal cylinder positioned horizontally on top of the starboard rail. It rolls freely and has perpendicular horns that keep the fishing line from slipping off each side of the cylinder.

They pulled until a couple of fathoms of line lay slack on deck. Kaare took the slack end and wrapped the line in the groove of the gurdy sheaves. The gurdy pulls the gear into the boat, using its sheaves—two metal discs the shape of dinner plates, placed bottom-to-bottom. The cast-iron sheaves are designed to grip a rope placed within the groove. If there is tension on the rope, the sheaves pinch the rope firmly—the more tension, the deeper it sits in the groove, the harder the gurdy can pull.

Freddy showed up on deck puffing three cigarettes that stuck out of his mouth and delivered two, shoving them between the lips of Wally and Chris.

Wally reached just under the railing, grabbed a lever, and threw it forward. Immediately the hydraulic-powered sheaves of the gurdy turned around and around, pulling on the line stuck in its groove. The gurdy complained, making a loud mechanical growl.

"Rraau-rrraau-rrraau-rrraau . . ."

Kaare sat down on a small seat to coil. The coiler is the human component of the longline hauling system—a crewman who coils the gear as it comes out the other side of the gurdy.

I would get to know the gurdy well, and I'd dread the way it forced me to work.

12

DEATH

RRAAU-RRRAAU-RRRAAU-RRRAAU . . .
Many tools of fishing exhibit shapes pleasing to the eye—the curve of
a hook, the spiral of a coil of line, the teardrop of a buoy bag. Different, to
the extent of not belonging, the gurdy looks ungainly. The *Grant*'s gurdy
was a hodgepodge of shapes joined together, built for function alone. It
stood on the deck and made a noise that grated on my nerves.

Rraau-rrraau-rrraau-rrraau . . .

The gurdy fed the longline to Kaare, who swung his hands back and
forth, left and right, grabbing the slack to make coils that fell from his
fingers, dropping perfectly to the deck in concentric circles. In just a
couple of minutes, coils of line accumulated between Kaare's splayed
legs into a cylindrical pile of rope a foot tall.

Chris stood next to the starboard side of the hatch cover in one of the
small areas cordoned off with boards. He was splitting up some half-
thawed "dogs," known as dog salmon or chum salmon. He threw slabs
of pink meat over toward Freddy, who hacked energetically with a heavy
cleaver at the fillets, whacking out serving-size portions on a wooden
chopping block.

Jack wasn't to be seen. I guessed that he was probably in the pilot-house or the engine room.

"Hey, Kid!" Kaare hollered over the noise. "So you know this thing is called a gurdy. Right?" he said, giving a nod toward it.

"Yeah," I replied.

"Do you know why it's called a gurdy?"

"Nope," I confessed.

"You ever heard of a 'hurdy gurdy' or a 'hurdy gurdy man'?"

I hadn't put it together, but when he said hurdy gurdy man, my mind started to scramble. I knew I'd heard the term before. Then it came to me—the hurdy gurdy was a hand organ. I recalled a poem about the hurdy gurdy man from when I was very young.

Kaare continued, telling me that the term "hurdy gurdy" comes from an instrument used long ago by street musicians. "Organ grinders . . . from Russia. . . . I'm pretty sure they were gypsies."

In my mind, I saw the hurdy gurdy man turning a handle stuck out the side of a music box. The box was mounted atop a short pole that allowed it to be rested on the ground. The gypsy worked the handle round and round, making music. From the poem, I knew there was a pet monkey involved, but I couldn't remember whether the monkey stood on the gypsy's shoulder or danced on top of the hurdy gurdy.

"It's a music box, Kaare," I blurted out. "I don't get it. This gurdy is hardly a music box."

"Before your time, sonny boy," Kaare went on. "There was a day that fishermen hauled longline gear up from the bottom of the ocean by the power of their hands and arms."

This didn't seem possible to me. I had an idea of the depth of the water here by how much buoyline had been used in setting the gear.

"Boats like the *Grant* here used to have a half-dozen dories stacked on the stern. They were basically sixteen-foot-long rowboats with oars and a little sail. Every day, the *Grant* would travel around the fishing grounds, launching the dories with two guys per boat. They'd set out a few skates of gear and let them soak for a little while. Then they'd take turns haul-ing them back. To haul the line back they used this thing called a gurdy. Nowadays we call the old-fashioned gurdy a hand gurdy. It got its name because cranking the handle looked like what the gypsy did with his music box."

I couldn't believe that a human could pull up a line that long—let alone from deep water with fish on the hooks, or an anchor.

What a crazy existence it must have been to be abandoned each day on the ocean in a little boat like that, I thought. The ocean was so big that the *Grant* felt small to me, yet I imagined being on the ocean in a dory. Starting at the bow of the *Grant*, I could walk sixty-eight feet to her stern. I looked around the *Grant* and began to appreciate how much room there was here.

Rraau-rrraau-rrraau-rrraau . . .

I would soon learn that the boarded off areas to Wally's right were called checkers. The two largest were for halibut—the live checker and the dressed checker—for live and dressed halibut. The two smaller checkers ran along the starboard side of the hatch combing. The forward one, the one closest to Wally, was called either the shack or the bait checker. The other smaller checker—three feet by a foot and a half—was known simply as the checker. This space was not for fish, but was essentially a hole to stand in. Chris, who stood inside it now, was working as the checkerman.

Rraau-rrraau-rrraau-rrraau . . .

Wally yelled out, "ANCHOR!" over the snarl of the gurdy.

Rraau-rrraau-rrraau-rrraau. . . . Clunk. It stopped.

Quiet.

The boat rolled gently to the port side. Thick jets of seawater gushed through the scuppers and washed across the deck, breaking the silence.

Wally bent his body in half over the rail, reaching headfirst. With a grunt, he reappeared, clutching a big fishing anchor and hefting it on board. From watching the gear being set, I knew that fishing hooks would soon follow the anchor. At any moment, we would catch our first fish. The prospect thrilled me.

Freddy appeared at the rail and took the anchor. Wally turned on the gurdy to haul in the next section of line slowly:

Rrr . . . aau . . . rrr . . . aau . . .

Freddy carried the anchor across the width of the deck, hooked over his shoulder, clearly laboring under its weight, with the strap of the anchor still attached to the buoyline. He moved slowly around Kaare to avoid walloping his stepfather's head with the flukes. He dropped the anchor with a clang to hang on the port rail. Freddy ripped a skatebottom free from a tied-up bundle lying on top of the hatch and flopped it down on the deck. He took the line from Kaare, sat down in the coiler seat, and began the process of laying coils down, this time on top of the skatebottom. Hooks were next.

With a *ting*, the first hook ricocheted off the roller's horn, falling to the deck on the end of its long leader—the gangion. Wally reached for the next gangion and separated it from the main line. He pulled hard to free the gangion, twisted tight around the main line. The hook at the end of the gangion whirled around the main line in a blur, spun free and was flung to the deck when coming clear.

The barren first hook disheartened me, and the second hook without a fish, even more so. By the third hook, I was questioning the overall efficacy of hook-and-line fishing, not to mention its place in the ocean, the judgment of my uncle, and his choice of a fishing spot. Hook after hook came aboard empty. What was going on? Where were the goddamned fish?

Wally tugged on the gangions, one after the other. Fred sat in the coiler seat, coiling the line as it spit out from the gurdy. Kaare stood next to the hatch cover, chopping bait.

. . . rraau-rrraau-rrraau-rrraau . . .

I looked around at the faces of the crew for any reaction to the empty hooks and found none. Finally, after several dozen hooks had been hauled, Wally caught a fish—an ugly bullhead-type thing, about a foot long.

Even across the width of the deck from Wally, I could see the fish flex its muscle, displaying a row of menacing spines sticking up in a line down its back and long spikes pointing out from all sides of its head. The mouth of the fish grimaced with what looked like a smirk. Wally ignored the fish's arsenal of spines and moved swiftly to position the crook of his gaff hook close to the fish's mouth. At the same time, he pulled hard on the gaff, twisting the hooked fish with a flick of his wrist. The fish spun off the line and shot, like a macabre projectile, directly into the bait checker, used for storing fish caught by accident. These would be sliced up and converted into bait—cod, bullheads, and the like. If the fish had screamed out loud in rage, the representation of what I saw would have been complete.

Suited up in oilskins, Jack approached. I asked him, "What in the heck was that?" I started moving toward the bait checker, drawn by curiosity.

"Watch yourself," he replied. "It's an Irish Lord. They've got a nasty poison in their spines."

I crawled over the top of the hatch cover, just behind Wally, and saw the fish lying in the corner of the checker. I could see why somebody might name this fish an Irish Lord. It had flared its fins and flattened its head as though furling its brow, looking very stern and haughty indeed.

I reached for a gaff hook and just as I turned the fish over on its back using the gaff's tip, "THWACK," another biological pincushion flew by my hand and pummeled into the side of the checker. Reflexively, I jerked my arm back. The gaff hook flew out of my hand and fell into the live checker with a loud clang.

"Fuck!" Wally cried, wheeling round at me, realizing that I had snuck into his working area.

"GODDAMMIT, Dean!" he screamed, with his face pinched. "Tell me when you're in there, for chrissakes!"

I looked to my uncle. Jack just smiled, chuckling to himself. His look had a way of making me feel more stupid. I retreated from the checker, sheepish, leaving the Irish Lord behind, a subject to be investigated at a later time.

When I neared Jack, he said, "You got off cheap that time. Listen up, now. You've got to look out for yourself out here. Use your head, else you'll get hurt."

I felt humiliated, and his "you-know-I'm-right" attitude only irritated me more. I resisted the urge to react. It was true—he was right. I didn't know much about being "out here." I began to unwind, choking on my pride with my awareness growing by a notch.

A minute later, I stood leaning against the rail next to Jack, still feeling the sting from my brush with the Irish Lord. Wally slammed the gurdy control with a smack to the "stop" position. Gaff hook in hand, he lurched over the rail, bent over fully, his knees braced against the rail and his butt sticking up in the air. He swung back with the gaff hook to strike. In a flash, he delivered the weapon down, over the side and out of my view.

Wally's knees slammed into the rail. Arching his back, he struggled to lift and rise back up to a standing position. He was nearly upright when his upper torso started to angle to his right—a fall, toppling into the live checker, was imminent. He paused for a moment, and then lifted with all his might, this time primarily with his arms. At the end of his gaff appeared the biggest fish that I had ever seen in my life. The giant fish balanced precariously at the very top of the rail, its body arched like a tightly strung bow. Ripples of muscle ran in a band of waves across its skin—smooth and pure white on one side. It was a halibut.

Very close to collapsing into the live checker now, Wally was overextending himself to lift the fish further. Chris turned away from cutting bait and saw Wally's predicament. Chris dropped his knife and snatched a spare gaff hook resting on top of the hatch cover. He spun around with the gaff hook raised like a fencing foil. With a snap of his wrist, he delivered the point of the gaff next to where Wally's disappeared into the head of the fish. Chris gave his gaff a jerk and the fish fell into the live checker. One hundred pounds of fury erupted. The fish began to dance within the confines of the live checker, alternately on its head and tail.

"Ah-right!" cheered Freddy. Kaare let out a whoop. With a wry smile, looking happy and little surprised, Jack chimed in, "Nice fish!" I was speechless.

Wally hauled the slack out of the line leading to the fish's mouth and grabbed a shorter gaff hook. With the same motion that he used to fling off the bullheads, he gave his wrist a quick flick and the big halibut fell to the bottom of the checker. Freed from the hook, the fish resumed its vain attempt at returning to the water. The force of the fish scared me.

Chris's eyes tracked the wild fish. He raised a stubby wooden club over his head, waiting for the right moment. The halibut flailed around the live checker, crashing into the sides, slamming the boards like they were sticks instead of heavy pieces of lumber.

"WHACK!" went the club.

All became quiet again, except for the rraau-rrraau-rrraau-rrraau of the gurdy, the tink of hooks, the mew of gulls, and the rush of seawater gushing in and out of the scuppers.

The fish had been killed by a single blow to its head.

13

LIVING WITH
RIGOR MORTIS

WITH THE DEATH OF THE FISH, AN EMPTINESS IMPLODED
onto the deck—for me, and for me alone. Looking around, I found the
mood of the crew unfazed and unaffected. Not so much as a pause in
their work. The vigorous meter of Kaare chopping bait thumped on,
pounding like a heartbeat.

For years, I had only encountered corpses of fish from the docks in
Seattle. This fish was the real thing. I yearned to see it alive again, hear
the wild flapping of its tail—the first live halibut that I had ever seen. I
fought with my feelings, struggling to come to grips with the rough edge
on the circle of life.

My grieving vanished in that instant when Wally slammed the gurdy
handle again, stopping the gurdy for a second time. He leaned over the
side and gaffed a lively halibut, albeit a smaller one. The fish shot out of
the water at the tip of Wally's gaff, sprung off the rail with a flip of its tail,
and flew into the live checker, flopping crazily. Its spirit restored mine.

"Keep it up, Wally," shouted Jack.

I joined in, giving Wally a cheer myself.

I sensed some irony in the way that Jack encouraged Wally, as if the captain was shifting the responsibility for catching fish onto Wally.

Chris heaved the big fish up to lie on the dressing mats on top of the hatchcover. He worked on it, removing the gills and innards. With a big knife, he slashed at the fish with a half-dozen broad strokes, jerked out a huge tangle of guts, and flung the bloody mass into the water. A flotilla of sea pigeons followed in the wake of the *Grant*. En masse, the mob attacked the blob of entrails, grousing and squabbling.

"Hey, Chris," said Jack. "Save the guts."

Chris just nodded, acknowledging him.

I wondered why he was supposed to save them. Yuck.

Chris stowed his knife by sliding it underneath the dressing mat under the fish. With a curious grimace on his face, Chris reached into the cavity with one hand, groping blindly. Finding his mark, he gave a yank and pulled out his hand, which hung onto two fleshy blobs, each the size of a Hacky Sack. Chris flung them overboard. Joined together, they orbited around each other and disappeared into the floating mat of feathers squawking next to the boat. By the way Chris tossed the blobs, it was obvious that he wasn't supposed to save them.

Chris reached for a strange tool with a rubber hose attached to one end and a short metal blade at the other, bent in half into a U shape. This was the scraper. Water spurted through the hose into the curve of the U, making the water fan out like a little toy water fountain.

Chris gritted his teeth, lifted open the belly of the fish, and began pulling the tool along the fish's exposed backbone, scraping madly. He attacked this job with ferocity. Bloody slurry splattered everywhere, pouring out of the gut cavity. I ducked away to keep from getting splashed. The scraper dragged across the backbone of the fish, making a noise like a stick across a washboard. Chris moved the focus of his work to the inside of the head with a dozen or so more strokes. Done with the scraper, he tossed it to the side to hang over a checkerboard. He spun the fish around and gave it a heave, sliding it to fall into an empty checker, aft of the hatch—the dressed checker. Fifteen seconds had elapsed since he started working on the fish. To me as a bystander, it seemed longer, like watching an eight-second bull ride at the rodeo.

Chris saw me staring. He dug his gloved fingers into the dressing mat where the fish had lain and scraped the blood and bits of minced fish toward himself, right onto his pant legs. At first, I thought that he was playing around. Then I realized that he was just cleaning up his work space.

"SKATE OF GEAR!" Wally yelled, shouting louder than seemed necessary, even when I considered the other noises on deck.

Kaare stopped chopping and hustled up to the coiler seat to take over coiling from Freddy, who by now had piled a tall stack of line, hooks, and gangions between his legs, higher than his knees. Freddy untied a knot, reached down and hefted his skate of gear up to his chest and tottered down the deck. He labored to carry the huge bundle of rope to the stern, paying attention to the heaving and rolling deck to negotiate the two steps that rose up to the level of the poop deck.

Kaare laid down the first coils of the next skate, and Jack started chopping bait at the block.

"Come here, Dean. I'll show you how to clean a fish," Chris said.

I looked over to Jack. He gestured, using the chopper in his hand, "Go on."

"Here . . ." said Chris. "Stand here, so you'll be safe from my knife."

Chris flopped a fish up to lie on the hatch, white side up. The dark side of the halibut—a deep emerald-green color—lay on the mat. Halibut, like all fish in the flounder family, have both their eyes on the dark side, which is the upper side of a flatfish when it is alive, swimming in the sea, or lying on the ocean bottom. Even though this fish was very dead, I was relieved that it was upside down, so I would escape having the lidless and lifeless eyes stare up at me.

With the knife raised in his right hand, Chris grabbed the fish by the gills with his left hand. He brought the knife to the throat of the fish and executed a series of slashing strokes. I had no clue what he was doing. In seconds, a jumble of bloody guts erupted out of the body. I winced at the dripping mass that Chris held in front of me, as if I was supposed to understand something. Still holding the gills, he flopped the guts down onto the dressing mat and scraped off an assortment of organs with his knife. He threw the stomach section, with gills attached, over to Jack. The rest of the guts went over the side into the water.

Jack chopped up the stomach, pushed the pieces into the bait basket, and grabbed a fillet of salmon.

I watched the pieces of gut get covered up by chunks of salmon falling from the chopping block. "You . . . have . . . got . . . to . . . be . . . kidding!" I blurted.

"Guts are one of the best baits for catching halibut," said Jack.

"Now that's what I call conservation," I laughed, amused by what I perceived as cannibalism.

When I turned back to Chris, his forearm and hand had disappeared inside the fish. He looked up and said, "Now . . . the balls."

He gave a yank and flaunted a pair of balls draped over his fingers. This set of organs looked much different than the Hacky Sacks of the other fish. They looked translucent and less rounded. Chris noticed my quizzical look.

"They're different," I said.

"It's a male," he replied. "Their balls look different. The other fish was a female and their balls are actually egg sacks. Until you open up a fish, you can't tell what sex it is, except that female halibut are usually much bigger than males."

This was really cool, I thought, despite my being an ego-challenged, ego-sensitive teenage male. This meant that halibut operated on a wholly different system than humans, where guys are usually bigger than girls. Under water, girls get to kick some ass. I cracked up and laughed, imagining a cartoon image of a hundred-pound female halibut having sex with a ten-pound male.

"You want to give it a try?" asked Chris.

"What?"

"Clean a fish."

I wavered, unsure.

"Here. It's not as hard as it looks."

Like most scenarios described as "it's-not-as-hard-as-it-looks," this was total bullshit. I trusted him anyway, because I couldn't see what he was doing while he cleaned the previous fish.

"Come on. I'll help you through each step."

To get his legs out of the hole of the checker, Chris raised his legs up one at a time, pulling his knees to his chest with his arms to clear each foot over the top of the boards.

"Oh, well. What the hell," I thought, and I climbed in, taking care stepping over to avoid cracking *my* balls on the crotch-high checkerboards. I slid the knife out from under the dressing mat and picked it up. It was huge—almost eighteen inches long—a veritable broadsword that made my pocketknife at home seem impotent.

Similar to a sword, the blade of the knife flexed some, a formidable tool. In my previous job, I had worked with a scoop at the ice cream store. Knives were much cooler.

Chris continued. "First of all, you have to gain access to what you are working on. Lift this gill flap and make a cut, so that the gills will open as far as possible."

I could see what I was doing, and I made the cut. Okay, that wasn't too hard.

"Next, here . . . grab the gills like this."

He reached under the gill flap and grabbed a handful of gills. He let go and I found the appropriate handhold, very firm and solid, made easier with the help of ridges of little spines in the gills that stuck into the fabric of my glove.

"Now, your next job is to get the gills free from the fish."

I had cleaned fish before—rainbow trout, little and delicate. To clean a trout was simple—slit the stomach, pinch the gills with fingertips, and pull—the guts and everything came out. Compared to trout, this halibut was a monstrosity.

"Here we go . . ."

Chris held his patience. He gave me instructions and encouraged me. He took over on a cut when I thrashed about. "Saw with a saw. Cut with a knife," he said. When I tried to pull out the guts while they were still firmly attached, he showed me where I had erred. Fish dissection wasn't coming naturally to me. I operated on a trial and error basis, which was very slow and tedious, with my knife passing blindly through flesh, constantly testing and exploring where the bones lay within the fish. This is the only way to learn how to clean a halibut, unfortunately.

"Saw with a saw. Cut with a knife." He said it over and over. It sounded straightforward and I tried to adhere to his advice, but without thinking, I would lapse into sawing with the knife as soon as I became tired or when I stalled halfway through a cut. Sawing helped me complete the cut I worked on, but it only made my work harder in the long run. Instead of working like Chris, who used one powerful slash with the razor-sharp knife, I sawed back and forth as many as ten times before completing a cut. I failed to make decisive thrusts, not possible with the weak muscles of my scrawny arms.

"See this here, way in the back . . ."

Everything looked the same to me—all blood and guts. Chris did his best to help me, and I did my best to comply. Nevertheless, it took me about ten minutes to accomplish what I had seen him do in fifteen seconds—to gut a fish. At last, and after a totally exhausting effort, my fish gave birth to a pile of guts. I felt wasted.

"Good job. That wasn't so bad," Chris said.

Was he kidding me?

"Now all you've got to do is get out the balls and scrape the bloodline."

From trout fishing, I knew that the bloodline of a fish was actually the kidney organ, which ran along the backbone inside the gut cavity. The bloodline of the halibut was huge by comparison. Every aspect of this halibut exceeded everything that I had ever experienced with a fish.

"First you have to cut the bloodline. I'll show you how to do it."

He took the knife and made a slit along each side of the membrane covering it.

"Now, this next part isn't so easy."

I thought, "Oh, great. It gets harder?" I began to lose what remained of my enthusiasm.

"I'll get the balls loose and then you can take over."

The skeletal structure of a halibut is simple compared to most fishes, as there are few bones in the body of a halibut aside from the elongated vertebral bones that pass through the fish laterally. But I didn't know that yet.

Up to this point, the gore hadn't gotten to me. My curiosity had prevailed—the same force that helps biology students explore the mystery of life through death, probing and prodding a frog's rubbery carcass.

"Here, let me have the fish."

Just as he had with the previous fish, Chris grimaced as he reached inside the gut cavity. I became tentative. What was bothering him when he did that?

"Okay. The balls are loose. You reach in and finish."

I cringed. Me? Reach in?

Deep inside my clothing, I felt little muscles tighten around *my* reproductive organs.

"Go with your fingers, in along the backbone, until you get to another bone that gets in the way. You'll find the balls just to each side of the bone."

At this point, my testicles raced to see which one could retreat into my body first. To quell a reflex that originated somewhere in my stomach, I gulped. I closed my eyes and I groped blindly into the cavity to find the backbone. Muttering to myself, I delved in.

"Okay. That wasn't too bad. . . . There, that's it. . . . There's the other bone . . . oooh! . . . oooh! . . . the balls!"

I opened my eyes and looked up to Chris, who clearly knew what I must do next. My voice thick with dread, I asked him, "What do I do now?"

"What the hell do you think! Pull goddammit. Go for it."

I gave the balls an unconvincing tug, and not convinced, they didn't budge. Dammit. I gave another tug. No response. I withdrew and tried to gain a better advantage, taking another grip with my finger on the lumps of flesh. I pulled again. Nothing. I started to get frustrated, which helped distract me from feeling nauseated. On the next try, I pulled so hard that I grunted. I felt the tissues holding the balls start to let go. Finally, after three more pulls, the balls came free.

Streams of sweat dribbled down my temples. What an ordeal. I felt like cheering out loud, like I had scored a touchdown. Entertaining my audience, which included Freddy and my uncle, I flung the balls overboard with panache. They smiled, amused by my incompetence.

After dispatching the balls, I thought that I was finished with my work on the fish. Jack, who coiled a skate now, could see this through my antics. He interrupted my celebration and piped in, "Hey, sonny boy. Pick up that scraper and get to work."

Recess was over.

I picked up the awkward tool and stared at it. Water burbled and spurted out the end. Looking closely, I could see that the U-shaped blade had been sharpened like a knife blade. To remove the bloodline, I mimicked Chris's work on the previous fish and started stroking the scraper along the backbone. The scraper shuddered down the row of vertebra. After a half-dozen strokes, Chris suggested that I take a look inside the fish to see what I had accomplished; I discovered that I had lots of work to do yet. Bits of bloodline were stuck between the vertebra and way back where the balls had been attached. I tried to hold the belly flap open to see what I scraped, but my hand and arm quickly tired. I had to resort to working blindly.

In two more minutes of scraping, I had made the inside of the fish clean and tidy.

"One more step, and you're done."

My hands felt like meat hooks. The muscle in the V between my thumb and forefinger was cramped so hard that I could barely hang onto the scraper.

Chris pulled the gill flap open.

"These are the "sweet meats," he said, pointing to some bits of darker meat near the head of the fish. "You have to scrape them out completely. The flesh of the sweet meat spoils quickly and will ruin the entire fish if it isn't removed."

Much to my relief, Chris grabbed the scraper out of my hand. He gave the fish a few powerful strokes, then opened up the gut cavity of the fish to show me the result. I was amazed. The water that spewed from the scraper had washed the inside of the fish. What had been a gross and disgusting mess had the appearance of being clean. The fish was now worthy of being presented to a cook.

"How about getting out of there. I'll take care of the other fish."

Wally had caught a few more halibut while I wrestled with my first victim. The fish lay scattered around on the deck at the bottom of the live checker. Exhausted, I crawled out of the checker and around the aft end of the hatch cover. Freddy stopped his pounding on the chopping block, taking care to avoid chopping off my hand while I passed by him. The club in Chris's hand rose and fell to stun another fish. For a while, nobody ordered me into another task. I braced myself against the rail to rest and to watch.

The deck had taken on the appearance of a factory assembly line, but instead of the crew maintaining stationary positions and working the same job over and over again, they moved around the deck and rotated through a series of tasks. Freddy had taken Jack's place in the coiling seat.

I hadn't expected Jack to work as a regular crewman, spending time on deck. I thought he'd stay in the pilothouse, running the boat. Through the open door of the bait tent, I saw him back on the stern, baiting gear. His arms and torso worked in rhythm, jerking from side to side, flipping coils of line down, alternately grabbing chunks of bait to spear onto bare hooks—throw the line down, bait the hook, throw the line down . . . throw . . . bait . . . throw . . . bait. . . .

"SKATE OF GEAR!" hollered Wally from the roller.

Now it became Kaare's turn to coil again. He stowed the chopper and grabbed a skatebottom. After the knot that tied the two skates together had passed through the sheaves of the gurdy, Kaare took over at coiling gear. Freddy untied the knot, stood up, and reached down to pick up his skate.

Some time later, I glanced back to the stern and saw Jack and Freddy, both baiting gear, jerking back and forth. They were out of sync with each other, marching to different drummers. I laughed at noticing that Jack's drummer beat his drum at a markedly slower pace than Freddy's. Instead of working at a disco beat like Freddy, Jack worked at the rhythm of a waltz. From his small pile of baited gear, I saw that Jack had a long way to go in order to finish his skate.

A few minutes later, having finished baiting—and lapping Jack in the process—Freddy returned to the main deck and began pounding away on the chopping block at fillets of salmon and bullhead.

"SKATE OF GEAR!"

Fred rotated back into the coiling seat. Kaare carried his skate to the stern.

After two hours of hopping from one job to another and watching the crew as they cycled through their rotation of jobs, I heard a heavy clunk from over at the roller. The noise shook the whole boat. Wally reached over the side and lifted the fishing anchor up and into the boat.

The first string had come aboard.

I felt like cheering but stopped myself, remembering that five more strings remained in the water, yet to be retrieved. I didn't have much energy left. My body felt drained.

I was doomed.

14

BEYOND
EXHAUSTION

WALLY PULLED THE BAG AND FLAG ABOARD. JACK HAD LEFT the deck and now steered the boat from the pilothouse. The *Grant* rolled back and forth in the lazy swell. Alternating from left to right, jets of water splooshed through the scuppers and washed across the deck. Instead of the growling gurdy, I heard the cackle of sea pigeons and mew of sea gulls. I relished the quiet of the moment. Fluttering in a light breeze, I spied a fleck of orange emerging through a curtain of gray fog. It was a flagpole that bobbed in the ocean ahead of the boat.

Wally untied the flag from the bag and handed the bag to me. Despite its balloonlike appearance, the bag was heavy. I fumbled, trying to heft it up onto the forward deck to be stowed.

Freddy popped up within the canopy of the companionway, startling me. He hollered out, "GRUB!" and disappeared back down the ladder. I looked around to the other guys for an indication of what I should do.

Preoccupied, Wally focused his attention on stowing the flagpole, pointing its tip at a square ring in the forward mast's rigging where the poles were stored. His target swayed with the roll of the boat, high above. He looked up and took a stab with the pole, looking as though he jousted, and missed. Before trying again, he paused and said, "Take the bag up to the bowdeck and tie it off."

The *Grant* moved faster now, into the waves, and the motion of the boat became exaggerated—the bow, more so. This was going to be tricky—scary, but exciting. I gave it a go, and as much as I tried, I couldn't keep my balance. I ended up crawling up the length of the bow deck and tied the bag to the pipe railing by its strap. I scuttled on all fours back down to the main deck next to Wally. I caught a whiff of fried bacon, its smell making me salivate.

"Good job. Now get your stuff off and go down and eat."

"Thanks, Wally. I'm starved."

I went to work at getting my gear off.

First down the ladder, I found breakfast foods spread over every available space in the galley. Scrambled eggs, fried potatoes, toast and bacon, butter and jam. Freddy stood next to the stove in a sweat-soaked T-shirt, smoking a cigarette. He watched over a half-dozen slices of bread that were toasting directly on the stovetop, bordering the hotpot nested within its hole, which began to steam a bit. My mouth watered and I struggled not to drool as I sat down. The boat creaked loudly, bouncing and protesting in the waves.

"Hey, watch out where you sit. When the boat's going up and down like this, you can get your ass pinched by the boards in the peak."

As I served myself sitting at the table, Freddy casually flicked a piece of toast toward me. It twirled like a Frisbee, landing in the middle of my plate.

"Don't be bashful. Eat all you can," he said. "It'll be a while before you get another meal." I shoveled food into my mouth.

One by one, the rest of the crew descended the ladder and filled in around the galley table. Wally was last, his cheeks fresh and reddened from his stint at the roller.

"Look at him, guys," said Freddy.

They all replied, either by grunting or nodding their heads.

I didn't know what to make of what Freddy said. Was it that I was eating too fast? Or, was it something else? I had assumed that table manners didn't count for much with this bunch.

Then Wally said, "You've still got your appetite. That's good."

Kaare nodded in agreement, "Keep this up, Slick, and you'll make it."

"I'm okay?" I asked, confused, with my mouth half full.

Kaare laughed. "Well, frankly, you are," he said, "Most people that go fishing for the first time puke their guts out by now. Some puke as soon as they step on a boat, whether it's rolling or not. The rest lose it when the boat gets into a swell. The rarest few, and it seems as though you are one of them, don't get sick. Sonny boy, the only test that remains for you is heavy weather."

"Weather?"

The others chuckled as Kaare laughed again. "The ocean's been as flat as a pancake since you've been out here." The smile left his face and he looked me squarely in the eye, saying, "If you can last through some nasty weather, you'll be doin' great."

My uncle came down the ladder and took his place at the corner of the table, nearest to the only exit.

For a moment, I scoped out my surroundings. I found the odors in the fo'c'sle appealing, with breakfast on the table and all. Truth of it was, I felt great—despite having the crew mentioning the idea of puking on a regular basis.

From the windowless fo'c'sle, a cramped enclosure mostly submerged into the ocean, I couldn't see outside, couldn't see the horizon. Its placement at the front of the boat exaggerated any motion, making the *Grant*'s heave and roll most evident down here. I knew enough about warding off seasickness to know that it was best to look to the horizon while on a moving boat.

I looked around. Drooped and strewn along the walls, spare clothing swayed to and fro, as though puppet strings controlled them. A pair of long underwear brushed at the back of my neck. Idle coffee cups hung overhead on hooks and rocked back and forth, appearing animated and not quite right; things looked nonsensical, like something out of *Alice in Wonderland*.

I resumed devouring my food.

The crew made quick work of the heap of food on the table. They ate fast, gobbling it down. As they finished, they started talking, telling stories, and commenting on the fishing—"not bad". . . "it's a scratch". . . "hope it holds up."

Jack was still working at eating his breakfast when Chris jumped up, grabbed the coffeepot from the stove, and started pouring cup after cup of coffee. The crew passed the emptied plates down the table to Freddy, standing at the sink. Packs of cigarettes popped out of pockets and appeared out of nooks and crannies. In seconds, cigarette smoke snuffed out the smells of breakfast.

Jack scooped the last bite from his plate with his fork, and before it reached his mouth, he said, "Okay, guys, let's get goin'."

I expected the crew to jump, but they just sat there, silent, coddling their coffee cups and staring into oblivion. In the morning, they had jerked into action in response to Jack's call. Now, it seemed they were ignoring him. Jack didn't seem affected by their reaction, or rather, their inaction. He stood up, put his plate in the sink, poured a hot cup of coffee for himself, and disappeared up the ladder. Not until he was gone, did the crew start moving.

The process of hauling the second string duplicated the first. The rest of the day dragged on, seeming to me an eternity.

* * * *

Midday, Jack brought the *Grant* up to the flagpole of the next string to haul. Chris stood directly in front of me, in the biggest open area of the deck.

"Watch this," he said, and put both his feet together until they touched.

"You'll be able to do this in a few days," he said. Typically, a fisherman's feet are spread out to a greater width than the shoulders.

He stood with his feet together and his arms out, keeping his balance on the rolling deck without holding onto anything. His body stayed perfectly upright, as if an internal gyroscope helped him to stand, incorporating gravity and the forces of the boat that affected him. According to Chris, I could do it someday. I was skeptical.

I mimicked his stance on the moving deck. Not able to anticipate the harmonic motion of the sea swell, I stumbled almost immediately.

"Don't worry. You'll get your sea legs soon," Chris said with a tone of confidence, making me feel better.

I had great balance, or so I thought. Why couldn't I duplicate Chris's stunt? This was absurd. I was a water skier and snow skier. Without hesitation, I could veer down a ski slope, bounce through mogul fields, fly over jumps, swoosh down icy gullies. But on the *Grant*, I looked like a staggering, stumbling drunk.

As we continued to haul gear, the boat and the waves worked together to make me weave and totter. I lost my balance a lot and with my hands occupied with fish and bait, I had little chance of catching myself. In every job on the boat, I worked with sharp objects—knives, hooks, and the chopper. My knees, hips, and shoulders took the brunt of the impact. Already, I had bruised myself on all sides from banging into checkerboards, railings, the hatch cover, and the bait table.

In my job at the ice cream shop, I had never been on my feet working for more than one shift—five hours, max. On land, standing up wasn't a task, but standing up on the *Grant* took a huge effort.

I was familiar with the term "sea legs," but I thought it meant something akin to a mental state, a level of professional competence that went along with being at sea; nothing to do with balance. I couldn't believe that I didn't have some inkling of sea legs in me with all the hours and days that I spent on the water on Lake Meridian. Only later would I understand that having sea legs is an adaptation of balance, taught by the ocean, to be learned by the body.

Late in the day, I was practicing standing up when I heard what would soon become for me a happy sound—a fishing anchor clunking against the side of the boat. Two more strings, and I would get to go to bed. I anticipated the moment when I heard the last anchor come aboard.

"Hey, Dino," said Kaare, "Before you hurt yourself, why don't you sit down and try coiling some buoyline."

"Sounds good," I replied. My feet ached from standing.

All day, I had secretly looked forward to trying to coil the gear. I had handled quite a bit of rope in my life already, through waterskiing, camping, and the like. I felt confident that I could coil buoyline. It looked simple, just like marching with your hands—left, right . . . left, right . . . left, right . . . repeat. Coiling the fishing gear, however, looked difficult because of the hooks attached to the line. I watched the others drop one hand away from the main line, then grab and swing the gangion and hook into the skate; it looked complicated.

Kaare sang out, "Hold 'er up, Fred." The gurdy came to a stop. It was Freddy's turn to be the rollerman.

I sat down on the coiler seat—a little wooden board with wooden sides to keep your butt from sliding off when the boat rolled.

"Go ahead, Fred," Kaare directed, "Easy now."

Freddy turned on the gurdy and the buoyline started coming in.

At once, the line fell off a pulley at the front of the gurdy, the fairlead. I hadn't been prepared to take in the slack.

"Hold it!" Kaare yelled out. "Okay, Dean. You've got to be on your toes when it starts. Pull against the gurdy and there won't be any chance of slack getting in the line and it falling off the fairlead." He tugged on an imaginary line to show me how.

When Freddy started the gurdy, this time I pulled hard. The line stayed on the fairlead.

I fought with the buoyline, trying to coil it down. The line was wiry and wanted to recoil like a spring. It didn't want to lie down flat on the deck.

"Hold it!"

The gear had fallen off the fairlead again. I had been focusing on the deck, trying to get the coils of rope to stay down and had forgotten to keep tension on the line.

"Go ahead."

I pulled harder. Kaare urged me on, coil by coil. My arms began to ache and cramp up.

"You can do it. Go on!"

Two minutes later, Freddy stopped the gurdy to haul the bag and flag aboard. I had managed to finish without a mishap, coiling 100 fathoms—600 feet. I was a wreck. I dripped in sweat inside my oilskins, like in a steam bath. I let go of the line and let my arms hang down at my sides apelike, trying to relieve the cramps in my arms and hands.

Kaare congratulated me, "Good job, good job," and whomped me on the back. His encouragement lifted my spirits and reassured me. I could tell that I had exceeded his expectations. Still I had many doubts about whether I was cut out for this profession.

*   *   *   *

The gurdy droned through nightfall. The fog pressed in closer; the swell was about six feet—moderate. The sea turned to black—invisible. I pretended that I traveled on a spaceship passing through a void without stars.

*   *   *   *

Deep into the night, toward the middle of the sixth and final string, I had met the demand of cutting and chopping bait for the rest of the fishing day. Chopping bait had been an unrelenting challenge put on me. It's an easy job for a greenhorn to learn. The rate of halibut coming aboard had slowed to a trickle. The checker had been emptied of fish. Nothing remained for me to do. Idle, I became sluggish and drowsy.

An inner voice took over my mind and spoke to me—coaching me like a director: "Okay, stay awake. . . . Just keep those eyes open. . . . Oops, watch your step. . . . Pay attention. . . . Keep your balance now. . . .Whoa, that was a big wave. . . . Be careful." Next, and worst of all, the "director's voice" helped me imagine that I was climbing into my cozy little bunk. "That's nice . . . ooooh, warm . . . and dry . . . and cozy. . . ."

A feeling of anxiousness came over me, and a rash, loud, and more authentic voice spoke out.

I don't like this! I feel like shit and I want to go to bed!

GODDAMMIT! WHEN WILL I GO TO BED?

Dreamlike words from another voice echoed into my consciousness, "Hey, Dean . . . take off your gear and hit the rack."

"Dean?"

Louder and sharper now, "DEAN!"

"Jeezus, are you awake?"

It was Jack.

"You're not looking so hot, kiddo. Before you fall flat on your face, why don't you go to bed?"

I turned and slogged up the deck like a walking corpse, weaving as I made my way toward the fo'c'sle companionway.

"No, Dean," Jack chuckled, "Turn around. Go back to the bait tent and take off your oilskins first."

Without as much as a pause in my gait, I wheeled around, back to the stern. Arriving in the bait tent, I woke up some and cursed bitterly out loud, realizing I had one more battle before I could go to sleep.

"These fuckin' things," I said, referring to my waterproof gear.

I got my jacket off fine, without too much effort, but my oilskin pants gave me grief. In vain, I tried to kick them off over my boots. Failing miserably, I took one foot and stood on one pant leg and tried to tear them off. They wouldn't budge. The wet, slimy pants seemed glued around the tops of my rubber boots. I had just wanted to go to sleep, now this.

I sat down exhausted on a damp skate of gear sprinkled with herring scales. I reached down and slowly peeled the legs of my pants over the boots, one at a time. Finally free, I hung the pants up in the bait tent. Nothing stood in my way. I was free to go to bed.

I made my way up to the fo'c'sle, walking up the deck. The crew still toiled, handling the gear and a couple of fish. A few muffled words of encouragement accompanied me as I departed, "Good job." . . . "Sleep well." . . . "Nighty night."

I retreated, crawling backwards down the fo'c'sle ladder, being careful. Normally, I went down the ladder facing away from it, using just my heels on the rungs.

I took off my boots and outer layer of clothing, climbed up and dove face first into my bunk. My whole body tingled, and my feet and hands throbbed. Two thin loins of muscle running along each side of my adolescent backbone, burned hot. Even so, I felt happy, relieved, and grateful, simply because I lay in my bunk, my refuge, my sanctuary.

Before I passed out, I recorded the moment, writing to Mom and my brother back at home:

July 8 - Got up at 4:00 AM Set out the line.
12-midnight—Still hauling set lines, worked my butt off. Got to go to sleep. Foggy and big rollers.

Four a.m. to midnight—a twenty-hour workday—my first day as a fisherman.

PART THREE

Adaptation

15

CIRCUMSTANCES

OUR CATCH OF HALIBUT ON THE FIRST DAY RANKED, ACCORD-
ing to the crew, as "just okay" and "nothing to get excited about." While
we slept the first night, Jack had made a two-hour move to the south, to
another spot farther offshore, attempting to find better fishing. All for
naught. Aspects of the first day overlapped to the next—fog, big rollers,
and few fish—blending time into a diurnal cycle of work far different
from my easy shifts at the ice cream store. By the end of the second day,
I had learned that discipline and repetition and routine structured my
new life.

Every day at first light, Jack steered the boat to where he believed
there were hungry halibut and set out the gear. The crew went through
their motions to make hooks and line fly out the back of the *Grant* and
settle to the seafloor. Jack set the gear in variations of two patterns: either
"end-for-end" in a linear series, or "berthed" side-by-side with a mile gap
in between parallel strings. Setting was finished by midmorning—sleep
time. A few hours of sleep—anywhere from four to six hours per day—
renewed us. Around noontime we woke, located the gear, and began the
process of hauling it back. Hauling twenty miles of gear took anywhere
from sixteen to twenty hours, depending on weather conditions, and

longer if the gear parted or broke during the day. The gurdy hauled, and the crew organized the gear and cleaned fish. At daybreak, the whole twenty miles of gear and 4,000 hooks went back into the ocean, unleashing another shitstorm of work for the next day.

After the first night, I received no more reprieves in going to bed early. For the rest of the summer, I stayed up with the rest of the crew through the night, until all of the gear had been hauled back. In the last half of the fishing day, sleep deprivation set in. I staggered around the deck, punch drunk, obsessed and anxious about my lack of sleep, in a state of nerves that escalated up to the moment I climbed into bed. I burrowed into my bunk at the end of the workday and slept like the dead. For each hour of rest that I got, I felt like I needed two. Each new day came too soon. A vicious cycle for a young body and mind.

Mimicking the lingo of the crew, I began to refer to midday as morning, because that's when we woke up. We ate breakfast after the first string, in the afternoon. Lunch was after the third string, late evening; and dinner after the fifth, after midnight. Just before going to bed at the end of the fishing day, in the early morning hours, we ate a light snack, called "mug-up," typically a steaming pot of canned soup. All said, "an average day of fishing" is an oxymoron.

The most extraordinary feature of working on the *Grant* was its degree of isolation. For the duration of the summer, I don't remember seeing any other boats or signs of humanity while we fished—not even a jet contrail in the sky. The *Grant* floated in the ocean all by itself, 200 miles away from Kodiak and 1,400 miles from its homeport of Ballard in Seattle—the farthest from home that I had ever been. We were completely removed from other people. For three weeks at a time, we became independent of civilization—inside a social envelope, hermetically sealed.

To sustain ourselves, we needed equipment and supplies. The boat provided everything—food and water, shelter and community. She supported us physically and essentially, not to mention her ability to float and keep us dry. Diesel in four large tanks fueled the engine, producing energy that propelled the boat toward fish and away from dangerous weather and shoals.

The irony of being surrounded by water and none of it to drink amused me. Being on the wave-tossed ocean was comparable to being in a desert, with salt water in the place of sand. Our modest supply of fresh water sloshed inside steel tanks underneath the floor of the galley.

Growing up in the Pacific Northwest, drenched by rain and surrounded by rivers, I had taken water for granted. Now, Freddy watched me and taught me ways to conserve water, and reprimanded me when I didn't, like using too much water to rinse the last bubbles of soap off of a cleaned plate. Freddy said, "You'll have hell to pay if we run out before the end of the trip." "If we run out of water," he added, "you'll be the one going down and chopping up some of the ice in the fishhold, so that we can melt it and use it for cooking and for making our coffee." I couldn't bear to think of eating noodles boiled in water contaminated by fish slime.

I conversed with only five other individuals for three weeks until we returned to town. Isolated socially and culturally, we had each other and nobody else. Jack, as the *Grant's* radio operator, was the exception. The long-range radio transceiver in the pilothouse served as our only link with the outside world. Jack used the radio only to gather weather information and to share fishing information with my other uncles. We received nothing in the way of news or current events.

Twice daily, the distorted voice of Peggy Dyson, the wife of a well-known Kodiak fisherman, Oscar Dyson, screeched out of the radio's little speaker. She recited "the Weather," as Jack called it, each morning and afternoon from her home in Kodiak to all corners of the North Pacific Ocean and Bering Sea. Peggy did this on her own and with her own radio.

Speaking into the microphone of her powerful radio set, Peggy transmitted over thousands of square miles to attentive fishermen and mariners who listened in and scribbled down the day's forecast on scraps of paper. Some recorded dismal statistics on coming storms and thereby wrote their own epitaphs, forecasts of approaching weather that would sink their boat and kill them.

The radio received other messages as well. Captains shuddered to hear their colleagues utter their last words through the tiny speaker of the ship's radio: "We're rolling over. . . . We're ro . . ."

Silence.

Every so often, Peggy brought cause for celebration over the radio waves, informing new fathers far from home of the birth of a baby, the condition of the mother, the child's weight, length, and sex. A glad thing. I believe that the toughest job Peggy placed on herself was telling fishermen that a loved one had died.

Our ship's radio was an unforgiving instrument, broadcasting messages of gift, as well as those of loss and wrath.

16

SURROUNDINGS

*"Drop your cocks
and grab your socks...
It's halibut hauling time."*

BY THE MORNING OF THE THIRD DAY OF FISHING, I LOATHED Jack's poem, finding it a horrible ritual associated with waking up. Being wakened by the shake of a leg, however, now offended me less. Within minutes of the crew's waking, a haze of cigarette smoke filled the confines of the fo'c'sle. Today, as I lay in my bunk, electric pain pierced my gut from a bladder stretched to the limit. I needed to get on deck and take a morning pee—and soon. The smoke added to my urgent need to escape.

"Hey. Look out. I'm coming down," I warned Chris, sitting below me on the galley bench.

I swung down from my bunk limber, slipped on my jacket and boots, and rushed out of the fo'c'sle, seeking to relieve myself and get some fresh air. Reaching the top rung of the ladder, I emerged into a shimmering panorama of the sea—a world without color. The sun's brilliance backlit the sky from behind a sheet of cloud, reflecting with such intensity that I couldn't keep my eyes open. I shielded them with my hands, like a prisoner rising from a dungeon. My eyes adjusted to the light and slowly widened.

Yesterday's fog had evaporated and I could see forever in all directions. A casual swell moved the ocean, its surface an elastic, ever-moving mirror. The air was still. As the boat rolled gently from side to side, the *Grant*'s guardrails sluiced into the water. Of the hundreds of sea gulls that floated about the boat, only two cried, taking turns bemoaning the absence of fish scraps to scavenge, waiting for the *Grant* to come to life.

While I absorbed the serenity around me, I stood at the rail to make my contribution to the salts of the sea. My gaze turned slowly from side to side, left to right . . . right to left . . . and without thinking, I began to look, searching for land. I scanned with increasing intensity. My focus became determined, converging onto a faint band just above the horizon—the narrowest of margins where the most distant objects could be found. But on this day, land was nowhere to be seen. I wanted to reject this idea.

I zipped up my pants and walked over to the starboard rail to inspect the 180 degrees of horizon that remained. But surely, I thought, part of the boat must have gotten in the way of my spying the coast. The outline of the pilothouse or the height of the bow must have obscured the shadowy shape of an island, certainly—or, if not an island, then perhaps a glimpse of the summit of a distant mountain, far away.

Jack had told me that we would likely get out of sight of land during the course of the fishing trip, but what he said hadn't sunk in. Something in me urgently desired to have solid earth materialize out of water and sky. Where was the land?

I stretched my neck out over the side of the boat in a futile effort to see a greater distance and searched again, giving up when I couldn't deny it any longer: we were out of sight of land. Muscles tightened and my mind began to race. My serenity vanished. A limitless expanse of water trapped me on the ocean, separating me from shore.

The roll of the *Grant* under my feet calmed my apprehension. My fear eased and I stood there in awe, face to face with this unearthly scene.

"Wait until I get back home and tell my friends!" my mind shouted inside of me. "This is so fucking cool. I'm in the middle of the ocean!"

But my euphoria waned. In relation to the ocean, I was insignificant. The *Grant* grew smaller, and with it, so did I.

Looking down to the water, I saw a ripple emanating from the waterline of the *Grant*. The boat moved forward through the ocean slowly. I assumed that the boat was adrift and out of gear. Odd, I thought, I couldn't feel or hear anything from the hull. I could usually sense the hull's characteristic vibration while under way. Besides, Jack sat at the table down in the fo'c'sle getting his coffee and a snack, not tending to steering the boat. This seemed to confirm my suspicions that the boat was out of gear.

Why was the boat moving through the water, or how?

Normally, a heavy chain stretched across the deck, leading up to the bow from the giant anchor winch on the aft end of the main deck, to the anchor on the *Grant*'s prow. The chain was big—the thickness of a man's arm. To keep it clear of our work on the main deck, the chain was usually hung up high on a large, steel cargo hook that dangled from the boom. But today, the chain was gone. In its place, a taut, rusty steel cable was strung low over the hatch cover, leading up to the bow, and disappearing over a roller where the anchor had once been.

We must be at anchor. Damn. How could that be? We were in the middle of the ocean. Converting fifty fathoms into feet, I raised my eyebrows at the result—the boat was anchored in 300 feet of water, secured to the earth.

I realized that it wasn't the boat that was moving—it was the water. But this was no river. The water traveling by the boat had no visible boundaries, no riverbank to contain the current or direct its flow. The whole ocean, as far as I could see, moved. My god, I thought. This is unbelievable. What power this current must hold.

I looked above to the clouded sky, spreading from horizon to horizon. Had I grown up in Kansas, this wouldn't have been notable. Back at home, our vistas included obstructions like trees, hills, and mountains. The smooth and featureless layer of clouds could have floated 500 feet above the ocean, or 5,000. No reference allowed an estimate. The cloud layer above appeared to bend over the rim of the horizon, seeming to form a dome or canopy, ever so slightly curved as it disappeared out of view.

The first ocean discoverers came to mind—the sailor explorers.

"The earth is round."

"No, the earth is flat."

From school, I knew that religion had played a heavy-handed role in this famous debate, long laid to rest. I'd learned that Galileo had challenged the Church, a giant body at the center of its own celestial system, stubbornly orbiting around its own ideals.

Discounting the value of hindsight, I stood upon the deck of the *Grant* and formed my own radical theory, right then and there, wagering that countless pre- and post-historic sailors understood that the earth was not flat. Perhaps they didn't go as far as sculpting a globe to fit their theories, but they definitely knew it was not flat. Even the most primitive sailor would have seen too much evidence, too many signs of curvature supporting the idea of an earth with a bend to it. With the dome of clouds above me, and harking back to the discussion I'd had with Jack about boats disappearing over the horizon only eight miles distant, I just smiled.

A muffled roar of laughter from deep within the fo'c'sle broke off my interlude with the universe. My stomach growled. I needed to get some food in my belly. I went below and wolfed down a bowl of oatmeal before going to work.

*   *   *   *

Just ten weeks prior, on April 27, 1972, Apollo 16 had returned from space. Images from the moon mission flashed into the nation's consciousness and inspired my curiosity. I rose above the drudgery of toiling on deck, levitating in the excitement of knowing that astronauts had once again explored and walked over the moon's surface. I came to believe that the passion of these adventurers evolved directly from the first ocean explorers. On the *Grant*, I learned to envy the old sailors in all ways. They had a world to discover.

17

THE CREW

IN MY FIRST DAYS AT SEA, MY NEW JOB HAD BOMBARDED MY
mind and body with information, new ways of understanding, and brutal work. I tried to keep my wits about me as I exhausted myself, learning
about fishing through a process of demolition and growth. I dreaded even
to move a muscle each morning, then appreciated the extra strength I
felt with work during the day. I observed a unique group of men and saw
how they interacted, in work and in play.

In the letter to my girlfriend, Patti, I recounted a scene from on deck:

July 10—2:00 AM–4:00 AM your time
Today, we had a damn funny conversation. It all started when we caught a
fish on a hook that was baited with halibut guts and Chris said, "I wouldn't
bite a hook if it had a piece of my fellow species on it." I couldn't resist but to
say, "What if you found a tit (female mammary gland) down there?"

Chris's comment, "Well, I don't know?"

Then Kaare started talking about the "tubes" on a dancer at Tony's (the
first time I ever heard a female mammary gland called that). I cracked up
after I heard that one. They keep telling me that I am going to get fixed up in
Kodiak, but I said I ain't getting fixed up with no fucking whore.

For the crew, talking about sex was the most popular recreational activity on board—by a long shot. The female body was an object of idolatry and discussing the sex act completely obsessed the crew. Except for my uncle, that is. My presence probably silenced him.

The crew's dialogue took a toll on my naivete. At fifteen, I had never had sex and never been around men who talked about having sex. In a few days I had gathered enough information and imagery in my head to write a screenplay for an X-rated movie—complete with choreography and lurid sound effects—and star in it. Not that I wanted to, that is, not right away. When I first stepped onto the *Grant*, I was a well-adjusted virgin. The more time I spent on the boat, however, the more anxious I became to lose that distinction and explore sex. But only under certain conditions and parameters—no "fucking whore" for me.

In the next installment of my letter to Patti, I gave a brief description of my workmates, crossing out the last line of the passage—but *without* making it unreadable. In my schoolwork, I normally scribbled over my mistakes, completely blotting them out.

Let me explain the crew,
Jack—my uncle, 30 to 40 years old,
cool guy.
Kaare—50 to 60, an old salt, dirty bastard.
Wally—30 to 40, nice guy, fun.
Chris—18 years old, 1st cousin-in-law,
quieter guy, dirty, is fun.
Freddy—30 to 40, dirty bastard, nice guy,
bawled 3 chicks in 24 hours.

By not completely crossing out the last line, it's likely that I was fishing for a reaction from Patti, saying, "Here's a guy who has had sex with three women in one day." Leaving it readable, I was posing ideas to her: "Hey . . . like maybe we can consider having sex together—just once, for the first time in our lives."

When I was a very young boy, my dad—bless his soul—tried to explain sexual intercourse to me. He explained "how the seed was passed from the man to the woman." That had confused me . . . damn . . . was it the seed? or the fertilizer? Anyway, I was confused. As a young boy, I thought that all people had penises, including women. Thinking that, it seemed awkward, if not impossible, to "pass a seed." Dad, who was by now in his third or fourth marriage (I never was sure how many) had a way of confusing me in his discussions about life.

My misconceptions (and my penis) were straightened out much later, through factoids in health class, talk among boys, and pictures in *Playboy* magazines filched from a friend's dad. But until now, I had never heard anybody talk about sex who had ever "done it." And these guys really knew what they were talking about.

"*Fuck*" and "*bawl*"? The latter term, I didn't even know how to spell.

In my youth, I was a linguistic chameleon. I enjoyed experimenting with the peculiarities of language, and cursing and swearing have always been the spice of marine jargon. True, most swearing can be boorish, but swearing with style is an art. Just like comedy, proper swearing requires skills of pitch and timing. Ignoring the vulgarity, I discovered that cursing was good, clean fun—and I had expert teachers.

For learning the more technical aspects of longline fishing, I had expert teachers as well. Imitating my shipmates was imperative and my habit of observing people benefited me greatly here. To succeed and to stay safe, I *had to* pay attention and copy the crew—I had no manual to follow in this job. The crew helped me when possible, but the operation of the deck was paramount and pulled them away from my lessons. By watching them, I noticed that at any indication of need at a working station, one of the crew materialized to fulfill the demand. How did they do that, I wondered. Sixty times a day, the rollerman cried out "SKATE OF GEAR!" and a crewman appeared at the coiler seat to take in the slack. As the crew flowed through their positions, moving from job to job, I found that if I looked at their faces, directly into their eyes, I could locate their focus. Through watching their eyes, I could see how they addressed the multiple challenges of working on deck—how they dealt with the flow of fish coming aboard and the long line that spit out of the gurdy's sheaves like an endless ticker tape. Through the crew themselves, I began to learn the rhythm of work on deck.

I bonded with Chris. At only eighteen, he was closest to my own age. I liked him as a teacher, too. Compared with the rest of the crew, he had recently worked his way through being a greenhorn and was still familiar with learning the tricks of the trade—like turning shortcuts into skills. Chris always responded to my mistakes or stupid questions with sympathy. I liked the way he corrected me with a husky chortle, "No, no, no

. . . . Saw with a saw. Cut with a knife." Through hard work, he had already earned the status of a full-share crewman. His salary equaled that of his seniors—Kaare, Wally, and Freddy.

Being taught by five people, each with a different method, presented problems. All too often I started a task, having just been told what to do, and the next person passing by corrected me. Although I followed the guidelines of my previous instructor, he had departed, leaving me to defend myself. As low man on the totem pole, I had to second-guess what five different people wanted. At times, it seemed that no matter what I did, it was wrong.

To confront this problem, I turned to Jack, the captain, for the final word on procedures and techniques. I soon realized, however, that he had a subordinate position on deck. Jack supervised the operation as a whole, but he left the details of the deck operations to Freddy and Wally, the leaders of the crew. They together shouldered the duty, each bearing the unofficial title of "deck boss."

Their personalities melded well. Freddy had the most experience of the two and was strongly intuitive. He could anticipate events. Wally excelled at solving problems and organizing the crew to respond. Freddy and Wally addressed problems cooperatively, as co-captains of the deck, bouncing ideas back and forth. For a greenhorn, serving two deck bosses could be difficult and frustrating. I wanted to satisfy them both.

In the varied jobs through which the crew rotated, each man had his own specialty. Kaare filled the jobs that never ceased. He kept to himself, chopping bait and coiling and baiting gear. At nearly sixty years of age, Kaare had more experience than anybody else on the *Grant*—even Jack. Nobody told Kaare what to do.

Because of Kaare's age, the crew excused him from the most strenuous jobs. He didn't go into the fishhold to ice the fish or to shovel ice, or to heave the frozen bait up from below. In recent years, he had also retired from the physically demanding job of rollerman, choosing to work only at the receiving end of the gurdy, coiling gear. When Kaare and I worked alone in the bait tent, talking together while we baited gear, he seemed a different man. Away from the others, he liked to bring up the topic of my schooling, asking me questions about my classes and other interests.

Years ago, Kaare had been a captain himself, of the longliner *Bonanza*, but he lost the boat to sinking. Freddy was also on the *Bonanza* when it sank, working for Kaare. When Freddy told me of a second boat—

the *Akutan*—that had sunk from underneath him, I wondered about his decision to stay in fishing.

Wally and Chris, as the principal workhorses of the crew, specialized in jobs of heavy labor. Their big bodies housed strong muscles, good for handling quantities of large fish—halibut. They excelled at dressing the fish and icing them in the hold, attacking these roles with warrior-like energy. Their power and size reminded me of a mighty statue that my grandmother pointed out to my brother and me as children during walks on the waterfront near Ballard. She told us that this man was Leif Erikson, a Viking who had explored the eastern coast of North America, 500 years before Columbus.

I once told Wally, "You look like a Viking."

He shot back, "I ain't no fuckin' squarehead," using the term that mocked the angular features of Scandinavians.

Freddy, of course, worked in the galley as the cook, but he also baited gear much faster than anyone else on the boat. He could bait a skate in just over ten minutes and frequently baited one-for-one—a rare accomplishment. The others generally worked at a pace of two-for-one, a speed at which one skate is baited in the time that it takes for two skates to be hauled aboard. Baiting at a two-for-one rate, or faster, was important, because the *Grant* had only two baiting stations. It was imperative that a crewman vacate his baiting station before the arrival of the next skate.

Compared to Freddy's speed, I was comatose. I wasn't fast enough to keep up with the turning gurdy and the rate of gear coming aboard—to bait two-for-one. The crew frequently gave me one of the last skates in a string to bait. That way, I had the most time to complete my job and not screw up the system. But even with the extra time of the transition between strings—anywhere from a half hour to forty-five minutes—I still toiled at the baiting station when the next skates came in. With their arms straining to hold the next skate—sixty pounds of ungainly rope coils—they waited to take over my baiting station. "Goddammit, get goin'," they barked. "Move your hands for chrissakes." And sometimes, "Get the fuck outta my way. Let me finish that goddamned thing for you."

Up to this point, I had made no real contribution to the *Grant*. I couldn't do much. The only job that I accomplished with competence was chopping bait. Even so, after an hour of chopping, my arm became paralyzed with fatigue. I didn't know how to coil gear or ice fish, and I slogged through futile attempts at dressing fish and baiting gear. After watching the rollerman go through his gyrations, I knew that a green-horn like me wouldn't be standing over at the roller for a long time—hauling gear and catching fish.

Daydreaming contributed to my sluggishness. I was homesick. I missed being on the Lake. I missed my brother and wondered what he was doing. Mostly, I missed my girlfriend Patti, and I thought about her a lot.

Because of my young age and extra need for sleep, Jack excused me from the daily duty of setting gear. This reprieve allowed me to sleep five to six hours—one more hour than the crew on average—a huge gift, indeed.

For the rest of the trip, only two men were required to stay on deck to set gear each day. The crew had split into two groups of two men each. Chris and Wally teamed together as one pair, and Kaare and Freddy as the other. Each pair missed out on two hours of sleep every other day. With Jack in the pilothouse during setting, that left one extra man—in this case, me, in my bunk.

In the halibut fleet, the cook was traditionally let off the hook for the duty of setting gear—which was, as far as I could tell, the only advantage of being a cook. But Jack had chosen to keep a sleep-deprived fifteen year old off the deck, away from hooks and gear hurtling into the water. Freddy took what should have been my position in setting.

Jack declared, "Dean . . . I hereby order you to be the galley slave. You will wash the dishes, clean the table, and dump the garbage." Turning to Freddy, he said, "If you need anything else done in the galley, you know who to go to."

Sounded like a good deal to me—and it was. I laughed to myself knowing that three times a day, washing the dishes would clean my hands. My mom would be pleased.

Instead of a noontime whistle to announce a break for a meal, Freddy's head popped up like a jack-in-the-box inside the companionway. He hollered, "GRUB!" and disappeared back down the ladder just as quickly. Sometimes, to continue the flow of work on deck, crewmen went down to eat, one person at a time, rotating through meals in the galley just like they rotated through the jobs on deck. I liked Freddy's cooking, and there was lots of it—piles of meat, potatoes, and gravy with all the fixins'. Breakfast was huge, too—hash browns, eggs, sausages, and bacon. But everyone on deck was hungry. Who would go down first?

Sometimes it was obvious. The crew had to coordinate. Whoever had been icing fish in the hold—the holdman—had first crack at eating. I watched Wally when he climbed out of the fishhold. He would be breathing hard, and his face dripped with sweat even though ice covered his oilskins. Someone would say, "Go ahead, Wally. You go first." The holdman got to take off his oilskins, sit down and cool down, and take a break. As I would soon discover, lifting and moving the halibut in the fishhold was like a vigorous gym workout with weights—big slippery weights.

The last guy left on deck at mealtime sliced up bait—octopus—considered to be the premier bait on the boat. Of all the baits used on the *Grant*, we cut the octopus bait last—just before we put it on the hooks. Octopus had the distinction of being the most difficult bait to handle. When it thawed out, the flesh became floppy—limp and flaccid—the same consistency as the live animal. Slightly frozen octopus was easier to hang on to and easier to cut because it didn't flatten under the pressure of a knife blade.

At the beginning of the trip, I asked Kaare, "What's the best bait?"

"That's tough to say," he answered, "because the bait doesn't necessarily stay on the hook after the fish is caught. You can't tell for sure. But even so, most halibut fishermen agree that octopus catches the most fish."

"So, why don't we put octopus on all the hooks?"

"We'd go broke if we did that, kiddo. It's very expensive. We dole it out to make it work right."

Of all the baits Jack used, octopus cost the most—$1.50 per pound. The cost of salmon varied between 50 and 75 cents per pound, depending on the timing of the annual run of salmon. Herring cost half what salmon did. Shack bait—the live fish that we caught besides the halibut—was free.

Jack watched me like a hawk when I cut up octopus, making sure that the weight and size of the pieces didn't vary—each piece needed to be about the shape and size of a deck of cards. Many times, he corrected me, "Those pieces are too thick!" or "Cut them shorter!" It surprised me that he wouldn't hesitate to criticize the others in the crew when he felt they cut the pieces too large. "Not so thick. We've got to make this stuff last." "Smaller!" he barked.

I had learned that costs related to bait were a crew expense, which meant that the bait bill was distributed among the crewmen. The *Grant's* crew talked a lot about "the expenses"—the money spent on bait, fuel, and food. They also spoke of the "Union" and the "Vessel Owners"—counterpart longline organizations founded more than fifty years earlier. Back then, the two groups had hammered out a contract still in effect, referred to on the Grant simply as "the Agreement."

According to the Agreement, I knew that Jack's individual share of the bait bill was insubstantial, so his attention to octopus surprised me. I wasn't aware of Jack's real concern, which was depleting our finite supply. If we ran out of bait out here on the ocean, purchasing more simply wasn't possible.

Aside from being a captain, Jack worked all jobs on deck. The roller was his forte. Jack worked with fast hands at the roller, dodging whirling hooks. He brandished the long gaff like a swordsman, displaying flair and élan in the art of catching fish. With the short gaff, he slung bullheads behind his back with a back shot into the bait checker. Of the entire crew, he was the master. He moved with precision in an exhibition of skill that excited me to watch. I could see in his face and through his body language that he loved this work.

In the 1960s Jack had worked for four winters as a king crab fisherman out of Kodiak and Dutch Harbor, supplementing his income during the off-season for halibut. In his last year of crabbing, he captained a boat.

I asked him why he didn't stick with it. He said, "I found crabbing dangerous and boring." Boring? I didn't fully understand, believing that *exhilaration* and danger went hand in hand.

Jack, Chris, Wally, and Freddy looked forward each day to their work at the roller, eagerly anticipating their turn as an opportunity to escape from the drudgery of baiting gear.

Rollerwork is complex and the rollerman must multitask: he must control the speed of the gurdy, gaff and haul the wild halibut aboard, clear the gangions, pull old bait off the hooks, and "shake," or remove incidentally caught fish from the hooks. Furthermore, the rollerman also controls the direction and speed of the boat. That's how Jack was able to work on deck and not be stuck in the pilothouse. The rollerman used a set of engine and rudder controls, positioned forward of the roller, to steer the boat over the invisible string of gear lying on the ocean's floor. Jack said, "I'd go crazy if I had to stand in the pilothouse all day, steering the boat."

With hooks flying around all the time, the rollerman's was a danger-ous job. Chris told me that fishing trips were commonly interrupted by hooksticks, or injuries from hooks. Fishing stopped while the injured crewman was taken to town for a doctor's help—to sew up flayed skin or to remove a hook sunk firmly into a hand.

On the *Grant*, hooks were a part of life. Handling the gear defined my day—hauling, coiling, catching, carrying, splicing, baiting, repairing, tying, stowing, and setting. Fishing gear represented everything we did. It measured time. Increments of hooks, skates, and strings gauged the day's progress, relating to the length of time that they took to be hauled aboard.

"After this skate, jump down in the fishhold and toss up ten salmon."

"I'll be over there to relieve you, in just a few more hooks."

"We'll have dinner in two strings."

"Let's give it a couple of skates before we cut up any more octopus."

"Only one more string, and it's time to hit the rack."

The motion of the gurdy's sheaves hauling the gear initiated the motion of the crew. When the gurdy turned, the boat worked. If the gurdy stopped, the potential for catching fish was put on hold. Time is money, and money came from a gurdy that turned.

Other than when we slept, the gurdy stopped for a hang-up when the gear got caught on the ocean floor, on something that wouldn't give. The rollerman had to stop the gurdy; if he didn't stop, the gurdy would pull the gear tighter and tighter until it snapped the line in two. Hang-ups happened more frequently when the gear lay on hard bottom—a rocky ocean floor—like off Foggy Cape, and the sharp pinnacles indicated by the ragged line I saw on the screen of Jack's fathometer.

Jack always took responsibility for trying to break a hang-up free. Before the line broke, or parted, he would excuse the current rollerman on duty. Then, using the gurdy to haul the gear tight, he would pull the boat hard against the hang-up. He steered the *Grant*, turning the boat around in big circles, trying to free the gear by pulling from all direc-tions. Jack chewed fiercely at the inside of his lip, while he tested the ten-sion of the line. The gear was stretched taut like a wire between the roller and gurdy. He pumped up and down with his hand on the line, testing the tension, making sure that it didn't become so tight as to break.

Sometimes he succeeded. Sometimes he didn't. A loud, "Fuck!" or "Shit!" accompanied the line's breaking. He'd throw his gloves down and whisk from the roller into the pilothouse, putting the boat in gear to travel down the length of the string and go pick up the remaining bag and flag. This was called "running the string down." More often than not, hauling the remaining gear went without a hitch. If the line broke twice during one string, we lost the middle, unretrieved section of gear—along with the fish hooked on it.

I dreaded the hang-ups. It took Jack a lot of time to deal with them. If the line broke, we had to run the string down once, to haul the remainder, and then again, to get to the next string in the series to haul. Both added hours to an already long day. If we parted the gear several times in one day, the day stretched into twenty-four hours or more.

For the most part, each day became more tolerable than the last. After I woke up and moved around some, I felt okay, ready for another eighteen hours or so on deck. Except for a few sores from pricks of fishhooks that had become infected, my hands remained in good shape, a remarkable achievement considering the artillery of weapons—hooks, knives, choppers, and gaff hooks—with which I worked. Best of all, I hadn't shown any sign of seasickness. My stomach hadn't given me any trouble at all. "So far, so good," said Jack.

Just days into the trip, I could tell that I was losing weight. The skin on my face felt tight over my cheekbones and the tiny deposit of fat that had once encircled my bellybutton had disappeared. I burned calories faster than I could replenish them, even though I ate like a starved animal at mealtime. The hard work and long days were taking its toll on me.

I thought about Mom.

Mom wouldn't like this, knowing that I was losing weight, though at the same time, she well understood that fishing was tough and that the job carried risks. I just had to be strong and stay out of harm's way.

After the fishing season one year, at a family holiday dinner when I was nine years old, my Uncle Donny turned up with a patch over his right eye. While fishing and hauling in the lines, a fishhook had sprung into his face like out of a slingshot and went into his eye, blinding it forever. Still, Mom's family had been lucky enough to have never been haunted by the sinking or loss of a boat or the loss of a crewman at sea.

Mom knew of the other risks associated with fishing. The crewmen with whom I worked were no saints and were well-known for their bad habits, abusing drugs, sex, and alcohol. She knew that I would be exposed to these people and their vices, and it was possible that I could be influenced by them.

Not yielding to fear, my mother had released me into a world where taking chances came with every waking day. She placed her trust in one beacon of leadership—her brother Jack.

18

MOM AND DAD

SUMMER BREAKS FROM SCHOOL STRETCHED MOM TO HER limit. She couldn't be at home to take care of me and my brother and go to work at the same time. Mom filled in the gaps by hiring a babysitter for us here and there and by sending us to a summer camp for a week— or two weeks, if she could afford it. Money was tight, yet Mom had a soft spot. She loaned my father money, which was seldom repaid. Always the salesman, Dad was great at selling Mom on his latest scheme and assuring her there'd be a bonus down the road.

My brother and I occasionally stayed with our friends' families, and Gramma watched us as much as she could. She was old, though, and couldn't handle two energetic boys for very long. Sometimes Dad helped out.

My brother and I were never sure, because it was always a surprise, but Dad would show up at the Cabin to take us away for the day. We'd go do something fun with him, like golf at the local course. I had learned the game quickly from him, most likely because Dad had a great swing and he was left-handed. With me being a "righty," I just could look at him and mimic the mirror image of his stroke. I'd hit the ball far and he'd say, "Nice shot, number-one son." I puffed up my little chest when he said that. I liked being his "number-one," being the oldest son and all. After golfing, he'd take me and my brother back home, and then he'd go away.

Two summers before going to Alaska, when I was thirteen, Mom let me go spend a whole month with Dad. We drove away from Seattle in his pickup truck, five hundred miles inland and north, up the Fraser River Canyon Highway. He was working up in British Columbia, selling rights to land he didn't own. Instead of Florida swampland, you got a great deal on an acre of forest where you could pitch your tent in the middle of nowhere.

I didn't know it when I left Seattle, nor did my mother, but Dad was having an affair in Canada with a local girl, just seventeen years old. I got to meet her. She was a sporty girl who wore tight jeans and cowboy boots, and her blond hair was cropped close. Though just four years of age separated me and my dad's girlfriend, she was already a sexually mature young woman and seemed a generation older. She was nice. Dad called her "Kickapoo," a nickname he used with affection.

One night in Canada, the three of us went to a barn dance at a nearby ranch. I got a big thrill out of sharing a bottle with Dad and Kickapoo, standing outside the big barn, taking straight shots of rye whisky before we joined the rowdy affair. It was the first time I'd been to a barn dance and the first time I'd ever quaffed a straight shot. Dad said, "After you swallow, just breathe out and everything will be just fine." He was right.

Most nights, Dad and I slept in a log cabin situated next to the shore of a remote mountain lake. Sometimes Kickapoo stayed, too. The cabin had one room, lacked electricity and running water, and used propane gas for light and cooking, as well as for powering the cabin's refrigerator—a curious thing to me. I never could figure out how that refrigerator worked—how a fire could cool things.

One night not long after the barn dance, my dad was holding a "business meeting" at the cabin, entertaining a couple of guys who sold property for him—his "Boys," he called them. He fed them rye whisky, the same stuff I'd tried, a brand called "Adams Private Stock." Dad enjoyed pointing out the connection between our family name and the name of the whisky.

As the night wore on, Dad and the Boys were getting pretty smashed. Kickapoo and I escaped the cabin to take a nighttime walk in the woods. We weaved through the black forest using the feeble beam of a flashlight to lead our way, shaking it to make the light turn on again when it blinked off. Kickapoo was good at listening and I did most of the talking, which was not like me.

When we got back to the cabin, the Boys were gone. My dad, quite unlike himself, lashed out at me, accusing me of making him look bad by leaving the cabin and going off with his girlfriend alone into the night. I didn't have a clue what he was talking about. And then it hit me, either by what he said next, or what I came to realize. He was ashamed of me because it looked like I had snuck off into the woods for a "tryst" with his girlfriend. I didn't even know what a tryst was. I still considered sex to be icky. When people kissed on TV, I closed my eyes.

I fell apart. I rushed outside to the porch, grabbed a hatchet, and started chopping kindling into pencil-size sticks. In the dark, it's a wonder that I didn't lop off a finger, or three, whacking away blindly with eyes filled with tears.

*   *   *   *

Mom never learned of what happened in Canada, but I believe she knew that I needed some extra help in growing up, to become an adult. Dad helped as best he could, teaching me ways to amuse myself as I grew up—playing golf, playing with girls, drinking straight shots.

I needed more. I needed someone to teach me how to work.

19

THE NIGHT

BY NIGHTFALL OF THE THIRD DAY THE GEAR HAD PARTED
twice, cementing the prospect of a long night ahead. At sunset, the sky
intensified, becoming a fountain of colors emerging from a blazing point
to soar overhead, sweeping over me and across the horizon in hues that
varied from the hot colors of a furnace near the sun to a cool, distant
lavender. Magnificent. I became awestruck, a mere speck on the ocean.

That evening, I initiated a ritual. Using up precious energy, I peeled
off my gloves, descended into the fo'c'sle, and groped through my stars-
and-stripes bag for my Instamatic. Back on deck, I raised the camera to
the darkening sky and, with the click of the shutter, tried to capture the
sunset. Time and again the resulting photographs were dull and grainy,
scant homage to the grandeur before me.

Sunsets ushered in the most challenging time of all—the Night. Night
and halibut fishing go hand in hand because of a hungry crustacean,
smaller than a maggot, that inhabits the ocean floor. During the day,
the sand flea hides in the bottom sediments. At night, it emerges by the
gazillions to devour anything made of meat, such as the bait and halibut
on our hooks.

The first time I saw them, I stood at the rail next to Jack, who was working at the roller. He gaffed and lifted a hollow halibut carcass, reduced to just skin and bones. Thousands of little creatures squirmed out of the fish's remains and dropped on the deck. They escaped the boat by slipping out the scuppers, awash in deck water. I shuddered at the sight.

"Oh, my god. What happened to that fish?" I was aghast. "What in the hell are they?"

"Biologists call them amphipods," shrugged Jack. "We call them sand fleas."

"I say they're a nightmare. What are they doing?"

"They're voracious eaters. A swarm can devour a fish in just a couple hours. They'd even eat you, young man . . . if you happened to sink to the bottom of the ocean."

A good reason to avoid falling overboard, I thought. A shiver crept up my spine. I scratched at an itch, fretting that a sand flea had been flung off the dead fish and gone crawling down the neck of my shirt.

Jack went on, "At night, they come out of hiding and become active, because they can swim and feed without being seen. If enough light were to filter through the ocean, they'd be seen by fish and crunched up like popcorn."

"That's one horror show I'd like to see," I answered, revolted.

Looking down into the water, Jack's eyes lit up. "We've got ourselves a fish, Dino! Grab one of those gaffs on the hatch and give me a hand."

I had looked forward to this part of my new job—catching a halibut.

I saw the halibut coming up through the water. It fought against the hook in its mouth and the pull of the gurdy. Its white side flashed below, illuminated by the roller light.

Jack turned to the coiler, shouting, "Hey, Chris! I need another man. We've got a wild one."

I turned and grabbed a two-foot-long gaff from a pile of a half dozen that sat on the forward end of the hatch cover. While Chris made his way around the hatch to join us, I went back to the side of the boat and stood elbow-to-elbow with Jack, who squinted, peering into the water to size up the fish. Estimating the size of the halibut was tricky. The uneven, restless surface of the ocean distorted its image. As the halibut came closer to the surface, its true size became more evident.

Jack hailed out, "Okay. I need everybody over here. We've got ourselves a soaker."

Chris turned round toward the stern and yelled, "We need a hand up here. . . . Everybody!"

I heard the rush of boots thumping on the deck as the guys working on the stern clomped up to the main deck. Raising their knees and legs over the checkerboards like runners clearing high hurdles, they grunted to get into the live checker with me. Gaffs clanged together as they were snatched up from the pile on the hatch cover. Each man turned toward the water, raising his rod of steel high in the air like a sword, keeping the curved tip out of danger from the others. We crowded the rail's edge, armed with weapons, ready to hunt, ready to attack. I loved it.

"Whaddaya think? Two . . . two-fifty?"

"Shit, I dunno."

"From here? Looks to me like closer to three."

This is so cool, I thought, realizing that halibut fishermen get to measure fish in increments of hundreds—*hundreds* of pounds.

Jack orchestrated the procedure, operating the gurdy like an oversize sport-fishing reel, playing the fish while taking care to not put too much strain on the gear, to avoid tearing the hook out of the fish's mouth.

Wally held a gaff that was almost four feet long—longer than anyone else's. He could reach the farthest and therefore he got the first crack at gaffing the fish.

Jack waited for the right moment, coordinating his timing with the roll of the boat when the rail was closest to the surface of the water. When the boat rolled away from the fish, the rail towered ten feet above the water. This put the fish far out of range of our gaffs. I could see that we needed to gaff the fish when the rail was closest to the water. This was a one-shot deal. If Wally didn't gaff the fish in the split second before the boat rolled back in the other direction, the gear would go tight and tear the hook out of the fish's mouth. The fish would swim free, with a rip in its mouth.

"Okay, here we go," Jack called. The boat rolled toward the fish. Jack turned on the gurdy to take the slack out of the line, coordinating the fish's arrival at the ocean's surface with the downward roll of the starboard side of the boat. It was time.

I expected Wally to swing the long gaff. Instead he plunged the end of the gaff into the water and pulled up with a jerk, snagging the underside of the fish's flattened head with the tip of the gaff. Wally controlled the fish now, pulling it within range of the two-foot-long gaffs that the rest of us held in waiting. The others' gaffs stabbed the fish. Thunk. Thunk. Tentative, I waited. The halibut was so big.

Jack called, "On the roll, guys." He added quickly, "Dean . . . gaff the fish, in the head. Quick."

"Now!" he yelled.

I swung, hitting the fish with another "thunk" and pulled with the rest. It seemed that the fish would never budge.

Jack directed us, "Ready! One . . . two . . . three."

The crew grunted in noises and curses, "Hhunnggg. . . . Uuoo-nnk. . . . SShheee-it."

We pulled in unison.

Jack strained to yell. "Come on guys! PULL GODDAMMIT!"

Grunting like weightlifters, we pulled, yanked, and pulled some more, using Olympian strength and determination to get the fish up over the rail and into the boat. At the apex of every roll of the boat, the crew gave a collective "uummph," inching the fish up the boat's side. Another roll, another grunt, another inch. The head of the fish crested the rail. Chris released his gaff, stepping up from the deck to stand high on top of the boat's railing. With the rest of us standing by, just hanging on, Chris teetered over the water and swung his gaff into the fish once more. With his back leaning into the boat, he tugged on the fish with all his might. With Chris pulling this way, we gained leverage. Soon the fish lay arched over the rail, ready to fall in the boat. I kept pulling, not realizing that my peers had fled the live checker like rats from a sinking ship.

"DEAN! . . . GET THE FUCK OUTTA THERE!"

I turned and jumped over the boards, narrowly avoiding getting squashed like a bug under the halibut. The 300-pound fish, 7 feet long, thundered into the checker.

"Yoo-hooo!" . . . "Yeah!" . . . "Nice one!" . . . "Good job, Jack!" The others cheered.

Jack just shook his head, looking at me with a sidelong glance.

*   *   *   *

The night wore on. Before sunrise, the sand fleas had eaten away at roughly one fish out of four, to the extent where they had no market value. When Jack and the crew returned the remains of the damaged fish to the ocean, they didn't react with a grimace or a word, making it seem that they accepted this loss as part of the cost of doing business.

Late into the night, I became aware that something had changed in the black sky. What was it? I leaned my head out over the side of the boat, peering out into the darkness. "Is the sun starting to come up already?" I asked myself. Squinting, I detected the slightest hint of watery gray light radiating through the black dome that had pressed down on the boat all night. "Oh, my god," I murmured under my breath. I had to look twice, because at first I couldn't believe it. Morning had come to a new day.

I hadn't yet learned that, as a halibut fisherman, I could expect to stay up through the entire night time after time and watch the sun rise, something I had done only once before as a kid at a summer camp.

We finished our work that morning under the brightening sky—with my horror of sand fleas serving only to tire me further. My muscles felt like cold lead. I entered a dreamy state of misery, exhausted, dizzy, depleted, more than I had ever been in my life. My body and mind told me that I had done something wrong. I felt as though I had slammed headlong into a wall. I can't remember whether I broke down and cried that particular morning, but if I did, it would have been at daybreak.

As we hauled the last of the gear aboard, I noticed that the *Grant* had begun to move with an exaggerated roll and bounce, agitated by wave upon wave of large swells. The waves moved the boat in a brusque manner, almost rudely, in much the same way that Jack shook me awake.

I thought the swells odd, considering that the wind remained light, just a little breeze, not enough to justify the height and mass of the waves that hurried past. The change made me uneasy. Wind and weather from somewhere else propelled these waves. My gut told me that tomorrow was going to be different.

Homesick, I wrote to Patti before dropping off to sleep:

July 11—Lonesome for you out here. Miss you.
We were around Chignik Bay, Castle Cape, Trinity Islands. Try looking that up in your Funk & Wagnalls Atlas. Now we are out of sight of land. They (the crew) are still on my back for the Kodiak bit. (Please pardon previous language.)

20

STORM DAY

FOR THE FIRST TIME WHILE AT SEA, NOBODY NEEDED TO shake my leg to wake me. The boat lurched up and down and from side to side, pitched by waves that slammed me back and forth against the sides of my narrow bunk. The crew's wool jackets, hanging from hooks on the walls, pivoted with the movement of the boat, swaying in unison like crazed, headless performers in a dance troupe. I asked myself, "Why am I here?"

A larger wave pounded into the side of the boat, plowing into the hull like a linebacker. The wave effectively tackled the *Grant* and pushed it over to lie on one side. I had played football in junior high school, as a skinny, second-string tailback, and I knew all too well the feeling of being hit by linebackers.

I slid across my mattress pad and grunted as my shoulder and hip smashed against the inner hull of the boat—one side of my bunk. The boat rose abruptly, then dropped in the opposite direction, angling steeply like it had fallen sideways into a gigantic hole. I careened against the outside rail of my bunk, raised only six inches higher than my mattress. I fell weightless, airborne in my bunk, for close to a second—a second that seemed like forever. The *Grant* shuddered as her hull plunged into the trough between waves, compressing my body solidly into the foam of the mattress.

If the boat dropped like that again, I might fly out of my bunk and crash into a heap on the galley table. Frantic, I curled up on my side and braced for the next wave. I extended my knees, shoulders, butt, and elbows out sideways to press against the inner hull and the railing, wedging my body into my bunk. I stopped sliding back and forth across the mattress. One helluva way to wake up, I thought.

I heard the sound of the engine slow and subside, then the pounding of boots across the deck. A wave thumped and burst, exploding against the boat's side. The shattering noise of shards of water followed, slashing against the pilothouse. Freddy climbed down the ladder. He turned around and saw me looking from my bunk. Water dripped from his hair and face. "That was a close one!" he said, raising his eyebrows and smirking like a child who had gotten away with something.

The engine revved up again, gaining speed, driving into the big waves, making the boat's movement become more aggravated. I guessed that Jack had slowed the boat to help Freddy traverse the deck safely—and almost dry.

During the foggy weather in the first days of the trip, Jack had said in passing, "Nice weather." He sounded sincere. Was he joking? But deadpan humor wasn't his style. Typically, there was a glint in his eye, or a sly grin.

I thought the weather those first days was horrible and I longed for sunshine. Everyone but fishermen would have considered the foggy weather bleak. The fog had blanketed the boat with visible water particles so dense that it accumulated on my face, forming droplets that dribbled off my nose and chin. Poor visibility gave us a lot of trouble, finding the next string to haul. But Jack *was* serious. For him, the calm gray days were nice weather.

The boat rolled hard to one side. A shaft of sunlight slanted through the companionway, slicing through the fo'c'sle like a laser. The sunny sky added insult to the irony, because for a fisherman, today's weather truly sucked.

Instead of being the last to rise, I was one of the first. Chris still lay on his back within his sleeping bag, staring upward at the bottom of the bunk above. Seeing me move, he stirred. Fred stood with his butt wedged against the sink. Positioned next to the stove, he was waiting for the coffee to boil in the hotpot.

From inside the pilothouse, Jack slowed the engine again. I heard another set of boots sprint across the deck. Kaare descended through the companionway. The engine sped up.

"It's nasty out there, boys," muttered Kaare.

I slipped on a layer of clothes and climbed up the ladder to take a look outside, protected within the shelter of the companionway. The companionway faced aft, away from the weather coming from the bow. White foam whipped from the crests of waves taller than the Cabin back home. I expected to see parallel lines of waves, like at a beach, waiting their turn to crash on the sand. Instead, the sea looked like one big river rapid—all chaos.

Hazy, the cloudless sky infused blue into the green ocean. The waves danced, alive. Sunlight pierced through the biggest waves, turning the water turquoise and lighting it from within. On deck, torrents of water gushed through the scuppers. Standing water rushed across the deck as the boat rolled. Spray hurtled over the bow and rails to splash loud against the pilothouse. A big wave smashed into the bow, filling the air with spray, and I lost sight of the pilothouse twenty feet away. A blast of saltwater mist covered me. I decided to retreat down into the fo'c'sle before I got too wet.

In the galley, everyone was moving slower than usual, settling around the galley table. The morning ritual never changed for the crew—cigarettes and coffee. I grabbed for a box of breakfast cereal and a bowl.

Wally snapped at me, grouchily, "If you've got an appetite in this weather, I think that you've passed the test."

"What test?" I replied defensively.

"What the fuck do you think?" he barked back. "Seasickness, for crying out loud." His edginess surprised me. "If you can hold your breakfast down today, I'd say you're one of the lucky ones. And believe me, consider yourself lucky if you don't get sick. It ain't pretty."

Nods of agreement from the crew.

"Sure as shit, we're not going to bring you in, unless you're throwing up blood."

More nods.

Motivation for not getting sick, I thought.

The boat slowed and the bucking motion eased. I heard the slam of the pilothouse door and a rush of boots. Jack arrived at the bottom of the ladder with rosy cheeks and a smile on his face.

"Good mornin', boys," he said in a hearty tone.

I thought, "Good, my ass," and considered Jack's potential for a corporate motivational speaker.

Freddy took Jack's cup from his hand and filled it with fresh coffee. On noticing me eating, Jack acted surprised. "Not sick, huh?"

Mouth full and chewing, I tipped my head to acknowledge him.

Jack tossed a sou'wester down next to me on the bench. "Here . . . you'll need this. Today, your baseball cap won't keep you dry." An odd-looking thing, a sou'wester has a bill that wraps all the way around, only a couple of inches in front but extending a foot long in the back.

Jack added, "It's good to see you eat."

The windowless fo'c'sle was turning in all directions—an ideal proving ground for seasickness. I didn't have any symptoms. I wasn't woozy. I felt alert and hungry. I had no headache, and my equilibrium was intact. In fact, moving around the fo'c'sle, my ability to keep my balance amazed me. I wondered if I had sea legs. I made a mental note that when I got on deck, I'd stand with my feet together and try to repeat Chris's demonstration from a few days ago.

"Did ya' haul the anchor, last night?" Wally asked Jack.

"Yeah. We were pulling pretty hard. It woke me up."

In the growing swells, the *Grant* had jerked hard on the anchor, which could easily have damaged the boat or broken the anchor gear. I wondered how I had stayed asleep through the clanging of the chain and anchor coming aboard.

Jack went on, "I hauled it. We drifted with the wind for about an hour while I grabbed the last part of a nap."

In that hour, the *Grant* had drifted several miles downwind from the end of our first string to haul. Now, we were running back to the gear, bucking into the weather—wildly. The autopilot steered the boat while Jack got his coffee.

*   *   *   *

That day in the storm, the *Grant* labored through the waves to move forward, forcing the rollerman to slow the gurdy down, turning at three-quarters speed. The boat couldn't keep up with the gurdy's ability to pull line on board. The line coming up, leading from the water to the boat, stretched forward and tight. Chris stopped the gurdy at times, to let the boat catch up with the gear. Halfway through the first string, I could tell that we had a long day ahead—longer than most. The novelty of watching the massive waves roll by wore off. The day turned into drudgery.

Early on, I baited a skate with Wally working behind me. The wind made the canvas of the bait tent slap hard and loud against its metal pipe frame. I yelled, "Wally, how hard is it blowing?"

"What?" he hollered back.

I yelled louder.

"How hard is it blowing?"

"I can't hear you."

I turned away from my skate and saw that he had his hood tied tight in a circle around his face. No wonder he couldn't hear me. I started to ask my question again. Seeing that he looked at my mouth, reading my lips, I lipped the words instead, "How hard is the wind blowing?"

Before he answered, the *Grant* came off the top of a wave and dove down, throwing me back into my skate. I caught myself before I fell, but I knocked over my skate in a mess onto the deck.

Wally just shook his head.

It took a moment to sink in. I had to start all over, picking up this snarl of hooks and rope—a quarter-mile long—and duplicate a half hour of work. Standing on the wildly moving deck doubled or tripled the effort it took me to do my work, and now I had to bait this skate a second time.

Later on, in the checker, more frustration awaited me. Interrupted in the midst of cleaning a fish to go help Jack pull a big fish into the boat, I left a 100-pound fish behind, lying on the hatch. The boat rolled hard to starboard. Unattended, the fish slid off the hatch cover and became wedged headfirst down into the checker where I had once stood.

The fish looked hideous, three feet of its tail sticking up, flopping back and forth to the *Grant's* roll. I finished helping Jack and crawled on top of the hatch cover, where I gaffed the fish's tail and pulled, angry at my mistake. The fish wouldn't budge. I changed tactics and pulled from a different direction so that I faced the wind. It was a vulnerable position and I knew it. A thick curtain of spray hurtled over the starboard rail and smacked me in the face. Salt water wicked into layers of clothing down through my collar and down my neck. At the top of my lungs, I cursed out loud, "FUCK!" only to startle myself—I had never said the word that way before.

I needed help. Nobody was available on the main deck. I scrambled on hands and knees over the hatch and went to the stern to ask Chris, working at the bait-table, to give me a hand.

Coming to the main deck, he chuckled and grabbed a gaff. "Good one. Happens to the best of us." With a short tug, he jerked the fish out of the checker singlehandedly. How could I be such a wimp, I wondered. Even with extra adrenaline pumping in me from anger, Chris's strength completely overshadowed mine.

The wind grew stronger as evening fell. The ocean's darkness absorbed the radiance of the light bulbs that illuminated our work area, so that I couldn't see a stone's throw beyond the railing. In the midst of a black void, the boat pitched, heaved, and snapped back and forth. Invisible, but audible above the shriek of the wind, I heard giant waves roar past through the gloom. The taut, thick cables aloft—the rigging that supported the masts—resonated in a deep hum, reverberating like the strings of a big bass instrument, making the deck tremble beneath my feet.

Through the darkness, I heard a deep roar. I saw Freddy's head jerk to look into the wind. A wave approached. We both saw the teeth of a white comber open, about to swallow the boat. Freddy slammed the gurdy off and screamed, "LOOK OUT!" and I squatted down into the checker and ducked. Detonating like a depth charge, the wave smashed against the planks of the old boat. Torrents of water cascaded over me and pounded against my back.

All work came to a halt. Tons of water slowly cleared the deck, flowing out the scuppers to return to the sea. Hunkered down, I stared out blankly from under my sou'wester, while remnants of the wave dripped in streams off the bill of my hat. A stream of seawater seeped into my undershirt, feeling like ice, having found its way through my raingear to soak several layers of clothing.

The crew called the largest waves "queer." For the rest of the night, the paths of queer waves crossed our position and unleashed their power against the side of the boat. About three minutes after the last collision, I would tighten up, unconsciously, with my muscles bracing for another blast. It always came.

I became accustomed to their timeliness but feared the possibility of even larger waves. I wondered if the ocean had the capacity to induce a wave that dwarfed the queer ones—a rogue wave. The uncertainty of darkness overwhelmed me.

Each queer wave wracked the boat like an earthquake. I recalled the time in my childhood when an earthquake jolted my home. That mild tremor made me insecure. I had trusted the integrity of my house without question—until then. That night on the *Grant*, I discovered that "seismic activity" is a joke to a mariner.

I hated the sou'wester. Its stitching leaked, and the rearward extension focused the cold spray and gale-force wind around my neck, creating a wind tunnel. I was chilled to the bone. Forks of pain coursed through my muscles. My fingers throbbed, swollen like sausages. The air was filled with ocean spray that dried on my face, crystallizing into a crust of salt that encircled my eyes. When a wave splashed onto my face, the crust dissolved into a hyper-saline solution that ran into my eyes, making them burn hot and weep. I was so hungry that I'd lost my appetite.

I grimaced with tears streaming down my cheeks, feeling completely and utterly defeated. To hell with it. I just wanted to quit, get off the boat, lie down, and go to sleep. But I couldn't leave. I was trapped, detached from everything I had ever known. What in the hell was I doing here? Does this make any sense, torturing myself on the sea?

I moaned and wished I could fly away. Held captive by misery, I let my thoughts take me far from the *Grant*, to another time and place. I left the ocean and retraced the events that had led up to my departure. In my daydream, I was with Patti . . .

*   *   *   *

Patti and I had said goodbye only six nights ago. It seemed more like six months. On the morning of Independence Day, we made a plan for spending our last evening together, watching the annual fireworks show from the shore of Lake Meridian. Among the kids who lived around the Lake, conventional wisdom had it that Lake Meridian had the best fireworks show in the state. I grew up believing that there was no better place to be. Nearing dusk, I waited for her to come to the Cabin.

I heard Patti's bare feet pad up to the doorstep. She knocked. I swung the door open and blushed upon seeing her. She wore swabby bellbottom jeans—the latest fad in teenager fashion. I wore them, too. But on top, a handkerchief-size swath of gauzy cotton veiled her chest. The white halter top clung to her breasts like electrified cellophane. Nipples pointed at me from beneath the thin fabric. She had never worn this top before—not around me.

Joined at the hips, Patti and I strolled from the Cabin down to the Lake, clinging to each other. We spent the night lying in the grass, kissing and mashing the button-fly of our jeans together, pausing only to breathe and watch the stars and explosions above us.

I'd been warned that by fishing offshore, away from land, I'd be giving up any chance of communicating with her for weeks at a time. No phone calls, no mail, no news. Nothing. I wondered how that would feel.

Lying next to her on the Fourth of July of 1972, I became distracted at times worrying about what I was getting into and what price I would pay for this opportunity to fish. Earning some money motivated me to leave. I wondered how much money I would make. That night by the Lake, I felt alone and I missed my girlfriend even before I had left her side.

* * * *

Now, more than a thousand miles away from home on a tempestuous black night, I wanted Patti. I wanted her comfort. I took a breath and resumed the job of gutting the fish that lay in front of me.

Tears of salt ran with those of heartache.

21

HOW TO JOG

INTO THE NIGHT, THE STORM GAINED MOMENTUM. WIND pressed on the backs of waves, transferring energy into the ocean, building reckless systems of liquid inertia to over twenty feet tall. Like a ponderous beast, the *Grant* lurched as though it tried to shake a clawed predator off its back. The crew rallied, stopping their work only when a wave breached the rail, to wait for the flood of seawater sloshing across the deck to clear out the scuppers. I had trouble standing up, let alone performing my job. Cloaked in misery, I felt pathetic—like a rag doll getting tossed around the deck.

Jack came up to me. By the look on his face, I could tell that he had something on his mind.

"I'm going to send you to bed, Dean. The weather's so bad that you could get hurt . . . you're a hazard to yourself . . . and the crew." What he said, I couldn't believe; I didn't challenge him. I had prayed to go to bed. "It's not just a problem of your getting hurt. If you got *seriously* hurt, we'd have to stop fishing and run to town." I imagined being injured and suffering through the 200-mile run back to Kodiak, to the nearest medical facility.

Much too rarely in life, a moment arises—"a dream come true"—this was one of those moments. I took off my oilskins, went below to the fo'c'sle, and burrowed into my bunk, grateful to be off the deck.

* * * *

When Jack woke me up later, I could tell that I'd slept for a while. Daylight streamed through the companionway into the fo'c'sle. "Go up to the pilothouse for a wheel," he ordered me. "You're going to take a jog turn." He turned away in a rush and scrambled up the ladder.

I lay in a crypt, deep within a lunging boat, squarely in the midst of a raging ocean and there wasn't a damn thing I could do about it. I couldn't escape. I felt worthless, and I thought about biding my time until I could jump ship.

I stirred. Pain coursed through my muscles and bones.

"Just get me through this trip," I moaned, resigned.

A swirl of cold wind snaked down through the companionway, brushing its dampness across my neck, feeling like clammy tentacles of an octopus. My body shook deep in a shiver. I raced to put on layers of clothes. Chris and Wally slept in peace, lying within their berths. Their big lumps of bodies lolled back and forth, like they were babies in cradles.

Now dressed and ready to make my way to the pilothouse for duty, I clambered up the steps of the ladder. Reaching the top, I halted and froze on gaining view of the biggest waves I had ever seen in my life. They were enormous, towering over the pilothouse. Millions of tons of water raced along the sides of the boat.

Getting to the pilothouse safely became my chief concern.

In the shelter of the companionway, I took time to pause and to scope it out. I estimated how fast I should race across the main deck, and most important, *when* I should start. I knew enough to know that if I ran in big strides like a sprinter, I'd crash. It wasn't safe. The *Grant's* deck rose and fell twenty feet. Running on a boat doesn't work.

When running on something solid like the earth, a terrestrial runner has faith that the earth will not move in the time it takes to complete a stride. If the ground were to suddenly rise by as little as six inches, a runner would trip and fall. If the land unexpectedly dropped, it would also make the landlubber stumble and fall. The *Grant's* deck didn't move in six-inch increments—it rose and fell twenty feet in a matter of seconds, not to mention pitch and tilt.

The threat of getting drenched complicated my plan. I wanted to stay dry. That meant avoiding getting splattered with spray, or worse—being exposed on deck and getting creamed by a wave. By watching the crew cross the deck in lesser weather, I had visualized their trick to staying dry. The *Grant* and the sea moved in cycles of differing periods. Their trick was to feel for the rhythm of the boat in relation to the rhythm of the passing waves. My challenge? Synchronize the two with the start of my dash, or else get drenched.

As the boat oscillated in relation to the ocean, it usually pitched down late compared to the next wave. The mass of the boat forced the bow to plunge deep into the ocean, sending either a burst of whitewater spray or solid green water flying over the bow. However, for fleeting moments, lulls in the two rhythms synchronized briefly, and they did so with a peculiar regularity.

I waited for the spray to clear the air and then tore across the deck like a bat out of hell. I scampered with my legs splayed out wider than my shoulders, taking quick, funny little steps—grace be damned.

I slammed the pilothouse door behind me, reporting for my first jog turn. I hadn't gotten very wet. I was spared.

Jack stared outside through the tiny windows. He looked strange. Coarse stubble peppered his face; the cowlick was gone. Sweat glossed his hair and matted it flat onto his skull. He wore a undershirt that hung limp from his scrappy frame. I was familiar with the Uncle Jack I saw at Christmas and Thanksgiving, cleaned up and dressed for the dinner table. This man was different—he looked withered and ragged.

He sat wedged onto the pilothouse seat in a calm pose, feet on the deck and knees locked like he leaned against a kitchen counter. The pilothouse seat was a slab of lumber the width of a single buttock. To keep from tipping, Jack raised his left arm perpendicular to his body and jammed his elbow firmly into the side of the depth-finder housing, cramming himself between it and the wall. The wind howled, whistling through cracks between the windows and their frames.

Jack's face was flushed red and he looked warm. I guessed that he had just come up from the engine room, having checked on various systems—the electrical system, bilge pumps, oil and water levels, and the gauges for the refrigeration system for cooling the fishhold. I'd been in the engine room a couple times. It sweltered like a sauna down there.

My uncle and I shared a space not much larger than a shower stall.

Immediately upon entering the pilothouse, I had noticed his smell. Like me, Jack hadn't bathed in a week and had just finished a long day on deck. He smelled bad. I wore a heavy jacket that trapped my body odor inside it.

I needed some fresh air and wanted to open a window. Even though opening a window was possible, it was not rational. All of the pilothouse windows faced forward—into the weather. Every ten seconds or so, seawater blasted against the face of the pilothouse, like a fire nozzle switched on and off.

Jack lit a cigarette. His fingers pinched and smeared the white paper with black grease. The smell of cigarette smoke became appealing to me.

Jack pointed to the scene outside and broke our silence, saying, "Whaddaya think?"

"wow!" I blurted out. Better words escaped me.

The *Grant* shook as a wave smashed the bow. A column of water broke free from the wave and hurtled airborne toward the pilothouse. It crashed against the windows, inches in front of me. I flinched.

Not breaking his gaze, Jack said, "You're going to have a jog turn." Still in the haze of waking up, I didn't bother to ask Jack what "jog turn" meant.

He had his eyes on something, but with the windows fogged up on the inside, it was hard for me to see out. On the outside of the windows, water droplets streamed along the panes, scuttling like bugs across the glass. I strained my eyes to see what he saw.

"There's one of our ends anchored out there . . . just a bag and flag . . . off the starboard bow."

"Your job for the next four hours is to keep it in sight and jog to it."

"Oh, great," I thought, wavering. As much as I tried, I couldn't see the bag and flag.

I couldn't hear the sound of the engine over the noise of the weather, but I saw that the boat moved ahead, in gear, heading into the waves at idle speed. Jack reached up and gave the wheel a quarter turn to starboard.

"To steer, you've got the wheel here," he said, touching one of its wooden spokes. "To take the boat in and out of gear, you have the clutch control for the engine." He reached up to the console to put the middle of his palm on a black plastic knob on the end of a short lever, and gave it a wiggle to show me that it moved back and forth.

Jack was proposing that I drive the boat all by myself—in a roaring gale. This might get exciting, I thought.

"It's tricky to keep track of the bag and flag," Jack went on. "The flag is all you can see at times, above the waves. Just stay close to the end, but not too close. You don't want to run it over and wrap the buoyline around the wheel—that is, the propeller."

"Where's the flag?" I asked.

"Over there, still on the starboard bow."

I tried to find a way to see through the spray smeared on the windows, rocking my head from left to right and back, like a predator trying to gauge distance to prey. I saw a distorted glimmer of red-orange—the flag.

"There it is," I cried out, relieved, seeing it flapping in the wind. In the next moment, I lost it again. Each time the flagpole dropped into the trough between the tall waves, it disappeared behind a wall of water.

"I've been jogging here for the last fifteen minutes and we're pretty much just holding our position at our present engine speed, but if you catch up to the end, take the boat out of gear like this."

He gave the lever of the clutch control a tug backwards. I felt the vibration of the engine subside as it went into neutral gear. Pushing the lever forward again, he put the boat back into forward gear, and the engine lugged down again.

"If you pull back further when you are in neutral gear, you will go into reverse gear. In this weather, if you get close to the end, just take the boat out of gear and you'll drift away from the bag and flag quickly. You shouldn't need reverse gear at all."

Jack took the last drag on the stub of his cigarette. Pushing against the force of the screeching wind, he cracked open the door and flicked the butt outside.

"You got it? Call me in four hours—at noon."

The bag and flag bobbed close by now.

"Yeah. I think I've got it."

A short course in jogging, indeed. Except for the part about keeping an eye on the bag and flag, it sounded pretty simple. I couldn't wait to take control of the wheel.

"The weather has backed off a bit, but if it gets worse, give me a call."

"Okay, no problem."

"Turn around every so often and take a look at the engine gauges. Right now they're in their normal positions. The engine rpm is set to idle at 650. You shouldn't have to touch it. The red handle next to the black handle of the clutch is the throttle control. Adjust it if necessary. If you need to speed up the boat while you are in gear, just bump the black handle with your hand like this." The engine rpm revved up. The thought of controlling the throttle of the *Grant's* behemoth engine made me feel giddy.

"All right now. I'm going to bed. Are you sure you've got it? Where's the flag?"

While he spoke, the boat had drifted away and I had lost sight of the flag again. My chest tightened with anxiety until I saw the flag about twenty degrees to the right of where I had expected it to be. It surfaced from behind a wave.

I exclaimed, "There!" expelling my breath. I took the wheel with both hands and gave it a turn to starboard.

"Be careful not to oversteer. The boat's heavy and it takes a while to respond to the rudder."

"Got it."

"Straighten out the rudder to the dead-ahead position before the bow swings all the way around. Here's the indicator for the rudder position right here," he said, pointing to a needle on a circular dial that hung from overhead. "Give me a call if you have any questions."

"Noon?"

"Right."

"Oh, yeah. If you do lose the bag and flag, sonny boy, it's your ass that stands on the bow in this weather to search for it, getting drenched with spray from head to toe."

He whisked into his stateroom, turning his shoulders sideways to fit through the doorway, shutting the door with a click. Jack rested in his bunk behind a one-inch-thick wooden wall just behind the pilothouse seat. Having him nearby reassured me.

Looking at the sea made me feel small again. I vowed to stay close to the flag and bag—at a safe distance.

Right away, I learned why the crew used bucking as the term to describe the pitching of a boat into waves. The boat felt like a bucking horse. The spokes of the wheel felt like the reins in my hands. When my stallion dove into the trunk of a wave, the beast reared back. A wild mane of froth flew back from the bow and crashed onto the waist of the boat.

Over the hours, the *Grant* never tamed. She bucked and shook, and she bucked some more.

I enjoyed my jog turn at first, fantasizing that I was the captain of my own ship, turning the spokes of the wheel, back and forth, left to right. I could see where varnish on the spokes had been rubbed away, worn by decades of fishermen tending to the wheel, polished by weathered skin. Oils from many hands, including the palms of my own grandfather, had stained the wood, deepening its color.

While I worked, the others slept. Even though I wasn't able to do a man's work on deck—that is, a full-share of work—I felt good knowing that I was contributing. I knew that Jack and the crew appreciated sleep more than anything else.

Before long I discovered that a jog turn was hard work, especially in a storm. Like a cat that batted a mouse, playing with it, the rough seas tossed the *Grant* back and forth. I fought to keep from being thrown against the walls. I found the little pilothouse seat useless. It was brutally uncomfortable. Standing up and hanging onto the wheel was wearing me out. To keep the bag and flag in sight, I had to press my nose right up to the windows. By the end of the first hour, I had exhausted myself, physically and mentally.

Jack was right. After turning the wheel, it took a long time for the *Grant* to begin turning and even more time to stop it. It took me a while to figure out that if the bow swung too fast, I had to countersteer. Extra work. I wasted a lot of energy, making up for mistakes. The hands on the ship's clock seemed to have slowed down, barely moving at all. I feared that they would never turn straight up to noon.

Finally, and precisely at noon, I opened the door to Jack's tiny stateroom.

His stateroom was even darker than the fo'c'sle, which was sunk deep within the *Grant*'s bow. This was Jack's personal cave. He had covered the porthole window with cardboard to keep out all the light. When he shut his door, the stateroom became pitch-black, even in broad daylight. The air inside was stuffy, too. The only ventilation for Jack's stateroom was a hole in the ceiling, just four inches in diameter. In contrast, the fo'c'sle had two sources of fresh air—a large funnel-shaped vent set into the bow deck, and the three- by four-foot opening of the companionway.

I felt sorry for Jack. I guessed how crappy I would feel after only four hours of sleep preceded by a long day of work. Despite how he woke me every day, I chose to shake Jack gently and not recite crass poetry to him. Nevertheless, I acknowledged feeling some power in being the one who woke him up.

In a minute, he joined me in the small space of the pilothouse. It surprised me how quickly he got up. He looked terrible, his face puffy and flushed bright red.

"Still got it?" he asked, rubbing his eyes.

"Yup," I said, with pride.

"Good. Good. I'm glad. Nice work. I hate looking for these things in this weather."

Then he looked me in the eye with a glare that said, "And don't you ever fuck up this job."

"I see that the weather hasn't come down yet," he said, turning to squint through the windows.

"Nope, hasn't changed."

My uncle took his place on the seat, rubbed his face and eyes some more, and stared out the windows, looking dazed. I tried to look vigilant, though I wished that he would take over the wheel as soon as possible.

Jack sat frozen in this position for a minute or so, and then his body jerked like somebody had poked him in the ribs. "Okay, listen up. . . . Call Chris and Wally on your way to bed. Tell them that I want to haul this end. They'll know what to do. You can also tell them that we won't be setting gear. We'll be running to Sand Point."

"Sand Point?"

"Yeah. We ruptured a hydraulic hose on the gurdy last night while you were sleeping. I jury rigged it, but in the long run I'd like to have it done up right. We can't fish in this weather, so we won't be wasting any time by going there. We can get new parts in Sand Point."

He reached up to take control of the wheel—finally.

I ducked out the door. Not caring whether I stayed dry or not, I rushed on my way up to the fo'c'sle. I fell asleep in minutes, well before Chris and Wally returned from hauling the end aboard.

Jack put the boat into gear, turned the wheel, and spun the *Grant* around to the west to head for Sand Point, traveling with the wind.

We cruised with a smooth ride in a following sea.

22

SAND POINT

SERENITY AND BLISS.

I woke up and thought I was still dreaming. The engine wasn't running. It was quiet. My eyes sprung open. "Shit!" I blurted out loud, and ejected myself from my bunk. We had arrived in Sand Point already.

"Those jerks. Dammit, anyway," I thought. I was pissed off that Jack and the crew had let me sleep. I urgently wanted to get off the boat. While I slept, I missed out on exploring Sand Point. Why didn't anyone wake me up, I wondered. I threw on some clothes, my heavy jacket, and my favorite hat—an English-style corduroy cap.

I climbed out of the fo'c'sle and nearly ran over Jack on deck. He dripped with sweat and his arms were black with grease up to the elbows. He'd been working in the engine room.

I confronted him, "What's going on? Jeezus, why didn't you get me up?"

He paused, taken aback by my aggressive nature. "I thought you needed the sleep."

"No way. I'll take any chance I can get . . . to get off this fuckin' scow."

Jack winced. "You're calling my boat . . . a . . . scow!?" he replied, feigning hurt feelings.

Blurring the line between sarcasm and truth, I exclaimed, "You're damned right it's a scow." We both knew that, among fishermen, "scow" was a mocking term for a shabby-looking boat. There is a type of boat called a "scow" used in Alaska, but it's basically a motorized barge.

Jack chuckled as I stormed past him. Yesterday's raunchy weather had left me feeling edgy. I wasn't in a mood to be polite or nice.

I clomped my feet, climbing up the ladder, and stood upright on the tall dock, which without warning seemed to sway under my feet. I became dizzy and struggled to keep my balance. Though the dock was firmly planted into the earth, rock solid, I had to shuffle my feet under me to keep standing. I'd heard that people felt this way after being at sea, but I had no idea that the feeling would be so strong.

The *Grant* was tied up in front of a fish plant—a cluster of buildings high on pilings above the harbor. It was a nice day with the sky gray and overcast. A light breeze scuffed the water of a channel that stretched a mile across to an island barren of trees and covered in grasses. I felt weird, being in a place without trees.

Jack yelled from the boat, "The rest of the guys went to the bar. We're leaving town in two hours. Don't be late."

Another wave of dizziness hit me. Not turning back, I took off down the boardwalk, walking with my feet out wider than normal to make sure that I didn't tip over. I wanted to find Chris.

As I crested a knoll next to the fish plant, wooden planks gave way to a dirt road. I gazed around looking at the landscape, acknowledging the distances I had covered since leaving home. I had started my fishing trip in Kodiak, 350 miles from where I stood. The cannery in Alitak was 100 miles closer. This place was different from Alitak. Alitak was solely an industrial outpost. Sand Point was a small village.

On a ridge to my right, I spotted a tiny building with a cross at the peak of the steep roof. Instead of one horizontal cross on the vertical, this one had three with the shorter bottom one tilted—the cross of the Russian Orthodox Church. A couple of stunted evergreen trees flanked the weather-beaten planks of the doorway. I took a position to frame the image with my camera's viewfinder and clicked a picture of church and trees, all hardened by the elements.

Behind the church, a stream ran through a gully to empty into the same bay in which the *Grant* floated. At the stream's mouth, two little wooden gillnet boats lay off kilter, half sunk in the water at half tide.

They were derelicts. Heavy scum stained what remained of the flaking paint, marking a brown ring around the boats' wheelhouses higher than the level of the water. I realized that the brown ring represented where the boats would be flooded at high tide. I wondered if these derelicts had been good boats in their time—reliable and seaworthy. Someone had taken care of them—long ago. I paused again to snap a picture.

Turning round, I saw Chris strolling down the dirt road and met up with him. He said he'd been on a walk himself. We headed up a dirt track, traversing a small ridge that divided the town in two.

"I'm heading for the bar, Dean."

"Me, too." I was excited to attempt getting into another drinking establishment. I'd tag along.

We passed through the door and saw Freddy, Kaare, and Wally in the bar sitting together. They looked relaxed, their beer bottles cupped in their hands. Unlike any of Kodiak's taverns I'd seen, this establishment had a row of windows that spread across the length of one wall, opening onto a sweeping view of the Shumagin Islands. I wanted to snap another picture but decided not to. I didn't want to draw any attention to myself.

The bartender swung around the bar, as Chris and I joined the others at their table.

"Hey there," he said. "Can I see your I.D.?"

I had my back to the bartender and jumped to the conclusion that he was talking to me. "That didn't take long," I thought. I moved to stand up and leave the bar when I realized that the bartender had directed his question to Chris, who fumbled through his pockets searching for his driver's license.

Unable to produce any I.D., Chris stood up and took off out the door without saying a word. As the door slammed behind him, Kaare chuckled. "Can you believe it? The punk kid got to stay and Chris gets kicked out." They let out a laugh. I blushed and scrunched down to hide behind the collar of my coat. Ten minutes later, Chris returned with his license in his hand and sat down. The bartender grabbed the card, scrutinizing it.

"Okay, fine. Whaddaya want?"

"Gimme a beer."

For the rest of our time in the bar, the conversation was dominated by barbs pointed at Chris. Wally and Freddy had no mercy. Given another chance, I wagered that Chris would have never come back to the bar. I didn't say a word or make a face. I just sat there and drank my beer, not wanting to hurt the feelings of the best ally I had in the crew.

*　*　*　*

We departed Sand Point on schedule and with a new hydraulic hose in place. Cruising out of the harbor to the north, the boat slid over the ripples that covered Unga Strait. Chris had climbed up and stood inside the *Grant*'s skiff, which sat in a cradle alongside the pilothouse roof. It took me several tries to heave the *Grant*'s heavy tie-up lines up to the skiff to be stowed.

Jack stretched his neck out the pilothouse window and puffed away on a cigarette. As he exhaled, the wind tore away at the smoke that collected in the eddy beside his head.

"Hey, listen up," Jack announced. "According to the broadcast, the weather forecast is good. . . . The outlook is good, too. We're going to run back to the east and set out the gear first thing in the morning."

All at once, the crew began to vanish like gophers diving into burrows below deck. Jack spied me trying to follow, but he wiggled his finger in my direction to come back. "Not you, sonny boy. You're going to have your first wheelwatch." I stopped, disappointed that I wasn't going to bed, but I realized at once that I would be driving the *Grant* while underway at full speed. I was thrilled.

At fifteen, I hadn't experienced the rush of driving a car on my own. To me, driving anything was a huge deal—a go-cart, a riding lawnmower, a motorized mini-bike—anything. I hopped into the pilothouse for duty.

Jack looked exhausted. "Now pay attention, dammit. I haven't slept much in the last forty-eight hours and I want to go to bed," he growled, flicking his cigarette butt out the window. "Once we get clear of the islands, you're going to be in charge. You'll be in charge of the boat and safety of everyone onboard. Our safety is in your hands, because the rest of us will be sleeping."

One by one, Jack pointed out the instruments and reminded me of their readings under normal operation: gauges for engine rpm, oil pressure, amp- and volt-meter, water temperature, and oil pressure for the engine and the drive transmission—the gear.

"The trickiest part here is the Iron Mike," he said, ". . . the autopilot. You need to learn some tricks to get it to work properly." He tapped on a black panel up by the compass. "In general, autopilots are known to be finicky. This one is pretty good. Watch what I do here." He reached down under the shelf that supported the compass and flipped a switch. Noise of an electric motor filled the pilothouse. The sound came through the planks of the pilothouse floor—from below in the engine room. The motor spun up to a high rate of revolution, whining loudly.

"To steer a course of northeast, or forty-five degrees, bring the boat around with the wheel, to steer that heading. Then push this small black button on the bulkhead to engage the autopilot." The electric motor groaned, sounding like it had taken a heavy load. It must be a tough job moving the rudder back and forth, I thought.

"Sometimes it works. Sometimes it doesn't engage at all. And then other times it holds a course, but it doesn't steer the right heading. It's a trial-and-error thing." Jack pointed to the compass. "See, we're steering forty-eight. You just keep trying until you get it right. If the wind is blowing the bow to one side, you may need to steer a few degrees into the wind, and then push the button. It should ease into a good course at that point. You try it."

I tried it a couple of times. The boat's course straightened, settling down to a heading close to forty-five degrees.

"Not bad. Keep your eye on the compass to make sure that we're following the right heading. Sometimes the autopilot disengages for no reason at all. Don't ask me why. It just does it. One more thing: when we're heading offshore, away from land and the danger of hitting anything, you can read a book while on wheelwatch."

"Are you sure, Jack?" It didn't seem right to drive a boat without constantly looking ahead.

"Yeah. Just look up in regular intervals to make sure nothing is in our path. There's a lot of room out here, and you'll rarely see a boat. We're moving pretty darned slow, too. You'd be hard pressed to find anything to hit." He went on to tell me about the risk of hitting drifting logs, and how to put the boat in neutral gear if we did hit one, but he added, "West of Kodiak there isn't one tree or forest until you get all the way to Japan."

He paused with a thought, then said, "However, today, I think it's a good idea if you just pay attention and not read until you get comfortable and get used to everything here."

Jack wrote out the names on a slip of paper. "Dean . . . Wally . . . Chris

. . . Kaare . . . Jack . . . Two Hours . . . 45°."

"You'll call Wally in two hours." He didn't hang around, but whisked through the door to his stateroom and went straight to bed. I was on my own—at the wheel of a seventy-ton boat. I wished that my friends could see me now.

I occupied myself by watching the compass and trying to understand how the Iron Mike worked. In light of the autopilot's idiosyncrasies, I could see that a wheel-watch involved a lot less effort than a jog turn. Instead of hanging on to the wheel, the autopilot did everything for me. I scanned the instruments and compass every so often and stared out the windows. I gazed at things outside the pilothouse—the sea and sky, and the birds. Over and over, I tried to find a comfortable position on the little seat, but failed.

My thoughts turned to worrying about the weather. The storm had taken a lot out of me. What would the weather be like for the remainder of the trip, I wondered. I had been on the ocean for only four days. I had eighteen days to go until the end of the trip. I felt good now, but could I keep it up? My transition from ice-cream scooper to fisherman had been swift and harsh. Though the rupture of the hydraulic hose was a nuisance for Jack, the intermission in the trip was a godsend for me. I desperately needed the break in Sand Point. The extra sleep was the most powerful healer of all.

I had a long way to go.

*　　*　　*　　*

We left the channel and rejoined the ocean. Clouds broke to make a patchwork of blue. The sea was peaceful now compared to the violence and noise of the storm. With not much to do, I was allowed a long, uninterrupted daydream.

23

LETTERS
TO HOME

ON THE *GRANT* I HAD PANORAMAS GALORE—360 DEGREE views, twenty-four hours a day. Within the tiny pilothouse of the *Grant*, wheelwatch duty granted me hours of daydreaming. I had much to process. I thought of home and family—and Patti. I knew that they worried about me and fretted over my condition. In a few days, my mom would receive the letter that I had sent from Sand Point covering the initiation to my new job.

My second day of fishing was summarized by an incomplete sentence.

9th of July—Chignik Bay area—Too tired, didn't write then, but am writing now.

That's all I wrote. I was finished after writing the word *"now."* I could have been implying a recovery from desperation, saying something like,

"Mom,
I'm still alive . . . I can write."

But it's more likely that I fell asleep with my pen in hand.

The next day's writing had more focus. I'd achieved some goals.

10th of July—Seeing a lot of those stupid sea lions (Jack got one the other day). I have been doing everything everybody else does, but "roller," "coil," "ice." I have been chopping bait for quite a while and I still have all nine fingers.

Animal life so far this trip—1,000,000 seagulls, 10 sea lions, 4 porpoise, 1 whale, otter and baby, 4 fur seals, 1,000,000 halibut (I wish), 3 King crab, 1 dogfish. Rollers, no sickness, partially cloudy.

"*Nine fingers*" is a joke directed at my mom. She fretted about my tendency to be accident prone with knives.

"*. . . those stupid sea lions, Jack got one the other day.*" Jack, like all fishermen of the time—longliners, seiners, gillnetters, trollers, and trawlers—carried a rifle onboard to shoot sea lions. Sea lions hung around fishing boats for food delivered to the surface by nets or longlines. Predators that competed too effectively with humans in the food chain were killed.

Sea lions loved to eat fish liver. I watched them chomp and tear open the bellies of fish with their lionlike jaws. After ripping out its liver, they let the fish go, alive and flopping, with guts and blood spewing from what used to be its stomach area.

Sometimes the halibut remained hooked to our gear after the attack. Pulled to the boat by the line, the eviscerated halibut was gaffed and pulled onboard, a hideous thing for me to watch. With horror and amazement, I gaped at the sea lions scavenging halibut. At the time, it seemed to me a good idea to react with the power of a rifle.

At night, Jack used sea-lion bombs to defend the gear from the sea lions' attacks. A sea-lion bomb is a big firecracker fitted with a waterproof fuse and a built-in pocket of sand to make it sink in the water before detonating. This allows the explosive force to stay in the ocean, rather than being wasted into the atmosphere.

July 12th—2:00 AM (4:00 AM your time)—We broke a hydraulic tube for the gurdy last night. We are going to Sand Point now. I might be able to mail this letter then if I can buy a stamp. I have gotten some neat pictures of some of the crew and surrounding islands.

Did some jogging for 4 hours. I might be coiling some skates pretty soon.

I wrote individual messages for each of my family, addressing my mother first. A week had passed since she had given up the captaincy of my life. I taunted and tormented her.

Mom—How many fingernails you got left?

I added a message to pass along to my father.

To Dad: Don't laugh too hard, but I got my 1st cut finger today . . . cutting a piece of bread for lunch. Don't worry . . . Jack was laughing as hard as you are, too.

Before licking the stamp and mailing the letter home, I added a footnote, in a scribble below "*Love, Dean.*"

P.S.—Just made it in a bar in Sand Point. Chris, my uncle's brother-in-law (the guy with the long red hair) didn't. (Real funny.)

In my letter to Patti I repeated telling the story from the bar. I added more in the closing of her letter, making mention of a secret plan and finishing with a forlorn goodbye.

When I come back (I am not going to tell you when) I'll rent a cab or something, or hitchhike (surprise you). I wish I could say more, but I miss you way up here.

[NOTE: In the year following my stint on the *Grant*, fishermen were forced to change their habits with the passage of the Endangered Species Act of 1973. Since then, the shooting of sea lions has been illegal in Alaskan waters. Convicted violators pay severe fines, in addition to serving jail time.]

24

CAUGHT
BY FISHING

FOR THE NEXT DAYS, I ENJOYED A PERIOD OF NICE WEATHER, which allowed me to gain strength. Short of praying out loud, I wished for calm seas every day. I worked, watching the transition from one day to the next, sky growing bright and falling dark.

I started another letter to Patti:

July 17 — 2:30AM — Can't wait to hit the bars of Kodiak. Really, I want to climb this big hill right by Kodiak and sleep overnight. There are some buildings up there from World War II. Me and Chris might make a carbide cannon when we climb the hill - and wake up Kodiak.

If there is anything you want to know about cleaning halibut, bullhead, cod, salmon, wolf eel, octopus, you know who to come to.

I am getting a bunch of names out here, "Speed," "Flash," "The Kid," "Dino," & "Slick."

As soon as I get paid regularly, I figure I will make $100 a day for 20 days (on a lousy trip) $300 a day on a real good trip.

Right now I may make $500 in 20 days.

Pardon my French, but fuck the ice cream shop. You would have to work 83 1/3 hours a day to equal this job.

In my letter to my family, I wrote:

July 14—12:30 AM—Yesterday, July 13th, we worked from 10:00 AM to 4:00 AM (the next day), so naturally, I didn't write.

Recently seen creatures are 1,000,000 gulls and sea pigeons, 10 wolf-eels, 2 sea lions, 1 skate (a small manta ray-like thing).

I COILED 2 SKATES. The skates were Canadian and have snaps on hooks, so it was like coiling buoyline.

One screwed-up day. Breakfast 6:00 PM, no lunch, dinner 2:00 AM.

July 17—2:30 AM—Working hard lately. We have 25,000 pounds total. Got to take a potshot at a sea-lion today with Jack's 30–30 carbine. He was 200 yards away and I missed.

Cloudy and calmer.

Two weeks into the trip marked a rite of passage—my sixteenth birthday. I became indebted to Freddy that day. He busted his ass just to make me a birthday cake—chocolate with chocolate frosting—my favorite. Because it was too hot in the fo'c'sle, he had to come on deck to "beat the mixture on high speed for two minutes" like the recipe instructed. The *Grant* didn't have electric receptacles. Fred beat it with a hand mixer.

When I went to bed the next morning, a dash of sarcasm lightened the reflection on my "sweet sixteen."

July 20—5:00 AM—Worked all my birthday. The gala celebration took place at Shumagin Flats with birthday dinner (beans and wieners) and a birthday cake baked by Freddy (real good!). I am reading the book "The Valachi Papers"—a Mafia book.

When the cake came out of the oven, Freddy told me, "You're lucky the weather is flat calm, 'cause I can't bake cakes when it's shitty out." That cake meant a lot to me.

* * * *

My writing ended here. Letters to both my family and Patti stopped. I abandoned reaching out to my girlfriend and family through words, in turn making my world collapse to become the deck of the *Grant*—17 feet wide by 68 feet long. By not writing, I had let go. Kodiak was still far away—in time and distance. Ten days remained in the trip.

Could I do this job? Something in me said "yes."

25

METAMORPHOSIS

FROM THE TIME I STEPPED ON BOARD THE *GRANT* IN KODIAK, the fishing life had swept me away. I was entranced by its wild nature—both animal and human. At first I resisted, hanging onto home and family and Patti, clinging to what I knew—my childhood. Swinging by a thread of stamina, I had surrendered and let go to become entangled, adapting to this difficult lifestyle and making it my own. Many factors challenged me seven days a week, twenty-four hours a day. I learned new ways to move, eat, and survive.

My greatest challenge was adapting to the motion of the sea. The boat's unceasing rocking and rolling handicapped me, the same as all fishermen. The rhythm of ocean and boat affected every aspect of my life and proved to be the primary obstacle to any task I faced.

Working on a boat forced me to live within a small space for every hour of the day, with no opportunity for relief or escape. All the halibut fishermen who worked in the schooner fleet shared similar living conditions, constrained by the boats' identical design. These vessels were not built for human comfort, they were built for catching fish. The largest space on board the schooners was reserved for dead fish. Fishing forces people to occupy a small space, which for a crew, instills proximity of body and mind.

All crews in the fleet shared habits that enabled success. These boats bound people together into a compact social situation, tied tightly within a culture of labor that practiced norms of behavior and tradition. Working next to my uncle and the crew, I learned about the conventions of the fishing life and conformed to certain practices and rituals—wayward habits as well as virtuous customs.

I was fascinated by the voodoo of fishermen and liked the way that their superstitions possessed meaning—both irrational and dead serious. I came to believe that certain actions tempted supernatural consequences. The superstition concerning whistling was my favorite. One day early in the trip, I innocently broke into a tune.

"Hold on for just a minute, kiddo. You can't whistle out here," Kaare corrected me.

"Why not?"

"You'll be whistling up a storm. You don't want to tempt a spell of weather to come visit us out here, now do ya?"

"Of course not."

Under those terms, I'd be crazy to whistle.

In the beginning, when I flipped the hatch over upside down to get into the fishhold, the crew barked at me, "Goddammit! Stop! If ya do that, the boat'll sink." They were serious. They scolded me many times until I learned better.

"Instead of lifting one edge of the hatch and flipping it over upside-down," they said, "just lift the hatch straight up when you set it aside." The crew explained to me that apart from keeping the hatch topside up, this extra effort also kept the boat from flipping over.

At first, I grumbled about the superstition regarding the opening of cans. Freddy stored our canned goods under the galley seats in dank lockers, where the moist, salt-laden air hastened the rusting of steel cans. It made good sense to me to apply the can opener to the cleanest and least corroded end of the can. "For god's sake, don't do that," Freddy told me. "If you open the can by its bottom, the boat'll capsize." I followed his advice and took no chances. After all, a little rust wasn't going to kill me.

After my first jog watch, I had mentioned to Chris that the boat felt like a bucking horse. He responded, telling me that I shouldn't talk about farm animals while on a boat. I never learned the reasoning behind that one. Even so, I respected the superstition as part of the nautical tradition.

At the time, I didn't think about whether these imaginary safeguards truly protected us from danger. I simply knew that through paying attention to these superstitions, I felt better about the boat not overturning. It was easy for me to see that superstitions give fishermen faith—that if we do certain things, we will be safe.

Through conversation with the crew, I learned the basic expectations of behavior—the unwritten rules. One unwritten rule stated that fistfights weren't allowed at sea. Fishermen were prohibited from resolving their disagreements by fighting—that is, not until the boat was docked in port. Chris said, "I've heard of guys who waited weeks for the trip to end, eager to get on a dock and pummel the crewman who had pissed him off."

This tradition stemmed from common sense. If someone got hurt in a fistfight while at sea, it could take days for the injured man to receive medical attention, like getting stitched up by a doctor. Not only would the injured man suffer in the delay, but the fishing operation would suffer, as well. Everyone on the boat paid dearly for a fight. Taking time off to travel to town cost the business many thousands of dollars.

Chris said, "Captains ask only one question after a fight: Who threw the first punch?" The violator stands to lose his job.

I don't know if it was a superstition, but no one on the *Grant* talked about procedures for accidents at sea—no emergency drills for fire, for a man falling overboard, or for abandoning ship. Nothing. I thought about accidents a lot, mostly about getting hooked and being pulled overboard.

I knew that we had an inflatable life raft on board, ready in case the boat sank. It was mounted on top of the pilothouse, stored inside a fiberglass case. But nobody talked about the raft, how to launch it, or how to inflate it. Likewise, no one told me if the *Grant* had a life preserver for me—not that a life preserver would actually preserve my life. I'd heard that the icy water would kill you. Even if you survived going overboard, the cold water froze your muscles, drowning you even if you wore a life jacket.

To heed superstitions can be similar to using good manners. Imagine being on a cruise ship and hearing a voice come over the ship's public address system: "Ladies and gentlemen, may I have your attention. . . . This evening, the movies *Das Boot* and *Sink the Bismarck* will be shown in the ship's theater. Thank you." With a click, the announcement ends.

Background music starts to play, and you recognize the theme song for the movie *Titanic*.

On a boat or a ship, this is bad manners. It simply wouldn't happen.

I had good manners. I had adapted to the constraints of living aboard the *Grant*, a smooth transition for me facilitated in part by having lived at home within the small space of the Cabin. I paid attention and picked up after myself at the table and kept my clothes stowed in my bunk and hung them on the hooks overhead. I worked harder to adapt to other, more basic things—like food.

Bread was one of my staples of life. I had to adjust to living without it. Had the galley freezer been bigger, Freddy could have frozen our bread in it, but the freezer was chock full, crammed with a three-week supply of meat. Halfway into the trip, our stock of bread on the *Grant* began to look like the mossy bricks for a garden footpath.

As much as I love bread, I cannot tolerate bread mold. I was not alone in this. One whiff of mold, and we tossed it out. No compromise or discussion. Into the slop bucket it went, no matter if we were down to our last loaf. About two weeks into the fishing trip, our supply ran out.

"Time to break out the pilot bread," Freddy declared.

"What's that?" I asked. I had my hopes up for something fresh.

"Here . . . I'll get some for ya." Freddy opened a cupboard door and reached deep into the food locker, where the dry goods were kept. The potatoes, onions, and canned goods were stored in the seat lockers under the galley table. The more delicate foods were stored in the standing lockers above. He pulled out a blue box with white lettering, roughly the size of a small loaf of bread. My mouth started watering.

"Here you go," said Freddy, tossing it before me. I became suspicious when the box hit the tabletop and made a thud like a block of wood. I tore open the package and broke open the clear plastic wrapper. It was filled with huge, round crackers, each the shape of an air-hockey puck.

I raised one up to my nose and sniffed. By now I had learned that before longline fishermen eat something new, they give it a good whiff and smell it. Like it or not, I was becoming a longliner. It smelled fresh.

I opened my mouth and chomped, crunching down to break off a piece. Crumbs sprayed from between my lips as I mumbled, "Damn. It's really hard."

Freddy laughed. "This is the stuff . . . like what Captain Cook fed his crew, along with worms crawling on salted beef."

"Oh, yeah. How old are these things? I suppose you've got some left-over salted beef from his expedition, as well," I replied. "What's this shit made of?"

"As far as I can tell, it's petrified wheat flour," said Freddy.

"You got that right."

For the remainder of the trip, I lived without bread and found that peanut butter and jelly on pilot bread sufficed as the best snack available.

I loved milk. At home, I drank nearly a gallon every day. My day started with milk poured over breakfast cereal and a big glass sitting in front of me on the table. But how long does milk last before it goes bad? I didn't know, mainly because my brother and I always drank it up before it spoiled. On a fishing boat, spoilage is a given. The question becomes, to what degree is the milk currently spoiled. In other words, how spoiled can milk become before a person rejects it? The answer? . . . It depends.

Milk spoils in stages. I got to study them all. The bacteria in the milk grows gradually but not steadily, and some cartons of milk spoiled faster than others. At the end, the process accelerated. The challenge was to train your nose to detect the difference between bad and intolerable. At home, I could smell the difference between milk in a carton just out of the grocery store and milk that had sat three days in our fridge—it had a slightly cheesy aroma.

Discussion and analysis accompanied every meal on the *Grant*. For the first test, we poured the milk in a mug.

"I don't know, Wally. What do you think?"

"Don't get that shit near me."

Wally had given up on experimenting with milk long ago.

"Here, let me see," said Chris. "Have you tried pouring it yet?"

"Yeah. No lumps."

If it didn't have lumps, someone would try it.

"Hell, it's not going to kill you," Chris said, ribbing Wally.

And it never did.

Following the habits of Scandinavians, we ate fish every other day. We ate fish at any meal, including breakfast. Initially, I found fish at breakfast odd, but I grew to like it.

Freddy prepared some unusual meals with fish. In addition to ordinary cuts like fillets and steaks, we ate cheeks, tips, and tongues.

The tip is the fleshy thing that extends into the space where a fish's gill flaps fold together. When a fish is cleaned, the tip is typically cut out with the gills and tossed back into the ocean. On occasion, Freddy asked the checkerman to cut tips for a meal. Tips are finger food. I learned how to hang onto the bone inside the tip and suck off the rich meat.

Tongues are not actually tongues. The tongue is a bite-size morsel of meat between the lower jawbones of a codfish. Roll the tongues in flour, salt and pepper, fry them up, and you have canapés for Norwegians.

I liked halibut cheeks the best. The meat of a cheek is very different from the rest of the halibut. I found them delicious, closer in flavor and texture to crabmeat.

To accompany a meal of fish, Freddy sometimes prepared sweet soup—a concoction that I loved. To start, he simmered a couple pounds of dried fruit in a pot of water sweetened with cups of sugar. The secret ingredient was sago, added to make the soup viscous. Sago is a lot like tapioca, except it colors the broth a deep red. At the end of the meal, Freddy ladled the soup into bowls and poured a small pool of cream on top. Sweet soup was warm and refreshing after eating a meal of fish.

On the days that we didn't eat fish, we ate slabs of beef and pork roasts with lots of potatoes. For breakfast the next day, Freddy chopped up the scraps from dinner and mixed them up with the eggs.

We had no space for ourselves outside of that used for sleeping and for food. The galley and the table—our dining room and kitchen—took up all of the floor space in the fo'c'sle. Our floating home had no room for chairs, dressers, couches, tables, or a bathtub. The only sink on the boat was the galley's sink. The *Grant* had no shower, not even a bathroom. Lack of plumbing didn't bother me, because personal hygiene didn't concern me much at all. For me, changing my underwear counted as hygiene. Kaare joked, "Fishermen brush their teeth once a week—whether they need it or not." Nobody on the boat shaved or combed his hair—except for Freddy. On warm, sunny days, he combed his hair back to keep the sweat from dripping onto his face while he baked in the galley, working next to the stove.

Instead of a toilet, the *Grant* had a "shitter"—a shoulder-width stall and toilet bowl incorporated into the port side of the pilothouse. Its door frame opened toward the baiting table, but the door frame had no door. It was a public place, devoid of privacy. I hated this. Most halibut schooners differed from the *Grant* in that their shitters faced aft, away from the bait table, with the only partners to your bowel movement being a flock of sea gulls flying behind the boat. On the *Grant*, the proximity of the defecating person to the bait table was such that the bare knees of the crewman relieving himself brushed against the wet, clammy oilskins of the person working at the bait table. I learned firsthand that when the defecator fouled the air with an offending smell, the tail of a foot-long herring would be slapped across the bare skin of one's leg by the baiter, leaving an imprint of fish slime and silvery scales. I deemed the arrangement of the *Grant's* shitter hideous, even for an all-male crew.

The timing of my bowel movements became paramount to me; I had to coordinate my "bathroom" visits with the bait table being vacated, which only happened for a few minutes between strings of gear. Of all the adjustments that I had to make, this felt the most awkward, not being able to sit on the toilet when my body said, "NOW!"

For recreation, I read books—primarily pulp fiction. At the end of each fishing day, I could only read for a few minutes without my eyelids drooping. After I tucked my book away, I fell asleep in seconds. All of the crew read books, and we talked about them while we worked on deck. Topping my list were novels concerning the Mafia.

Unlike the rest of the crew, I wrote, recording my experiences, thoughts, and feelings in letters to home. During the first part of the trip, my need to write was compelling. I was overwhelmed by my experiences and needed to share them somehow.

While working on deck, I daydreamed about writing, laughing to think about composing my essay in the coming school year—"What I Did on My Summer Vacation." Next year's ice breaker in English class would never be the same. In my daydream, I imagined my teacher asking, "Dean, would you please stand up and tell us about your essay? What did you do on your summer vacation?"

"Let's see, I worked on my Uncle Jack's boat in Alaska . . . with him and his fishing crew."

"Oh, yes. And can you tell us what you did with your uncle and his fishing crew?"

"Ummhh. First, we visited the strip club in Kodiak and I ordered a beer . . . and then the table-dancer got right up in front of Freddy and . . ."

My childhood run of summer vacations had come to an end. Giving up "my summer" was a huge loss for me; it had been the best season of all. My favorite times were all from summer vacation . . . the road trips . . . summer camps . . . swimming . . . playing ball . . . hiking . . . and that special trip to Disneyland. After school, after nine months of academic toil, I felt entitled to *my* summer vacation. I became insufferable if I wasn't outside. I played, and then I played some more. I didn't "work" on my suntan; I played, becoming bronzed in the process.

I had reached an inevitable transition, forcing me to grow up and go to work. Fishing had its trade-offs, both advantages and handicaps. The upshot? I was on a huge adventure. The downside? Isolation.

Unlike my friends' summertime jobs at home, my adventure kept me from playing after work, waterskiing and swimming at the lake, and from generally being a kid and just screwing around. The normal sixteen year old eased into the adult world of occupation, taking on simple jobs with simple policies and roles. The job description in a stereotypical job for a teenager—mowing lawns or gardening, for example—reads: A. Remove weed. B. Put weed in wheelbarrow. C. Insert dollar into wallet.

This was not my fate. I worked in a precarious place, far from home. I was outside, but trapped. My prison yard, the deck of the *Grant*. . . . My penitentiary, the ocean. . . . My respite, drinking a beer in a strip club. Altogether, these were extraordinary settings for the essay: "What I Did on My Summer Vacation."

26

PERSONAL
FACTORS

AS THE *GRANT*'S CAPTAIN, JACK FILLED THE LEADING ROLE on the boat, and the rest of the guys embodied the crew. I wasn't one of them as yet. I was still the greenhorn, and I felt like a third wheel. My status wouldn't change until the day I became a full-share fisherman. I had a long way to go. My physical strength was hardly that of a man. To earn a full-share position on a longline boat, it takes a grown man and a full season of fishing—about six months of work and training. I was not a grown man.

In the period encompassing my sixteenth birthday, I had evolved from a kid scooping ice cream, working with my friends and peers, into a young man working among men. The crew of the *Grant* took care of me and began to fulfill my needs in a working community—supporting and feeding me as the weeks progressed. Together, we became a hairy and ragtag bunch, personifying a motley crew. I had a sense of belonging.

For me, a combination of factors began to align like pieces in a puzzle. To succeed in fishing, I must have all the pieces and must fit them into place. Already, I had dealt with the question of seasickness, and my sea legs had developed to the point where I could safely withstand working on deck. Instead of jumping, using no hands, off something like the hatch or forward deck, I now used my hands to ease myself down. I moved around the deck now with the sureness and agility of a genuine fisherman. Down in the confines of the fishhold, I had to face claustrophobia. I squeezed through tight places while sliding on my belly and learned how to work while hunched down on my knees. Aside from perpetual weariness, I felt fine. I had begun to cope with sleep deprivation. This was key, the same as dealing with the carping of the crew. I couldn't afford to burden myself with their complaints, allowing my emotions to be worn down. I came to know the difference between helpful criticism and ornery nagging, and to sort out what mattered.

I had learned that I could deal with the intimidating power of the sea—the storms, the fear of drowning, and the fear of death. As I began to explore my abilities, as well as my weaknesses, I found that I possessed the basic requirements for being a fisherman. I began to feel like I was part of an alliance—and part of a team. I felt good and my body was holding up. My confidence grew.

At the beginning of every day, I endured the agony of waking up. Waking counted as my day's first accomplishment. Time after time, I suffered through my uncle's raunchy poem—my first audible impression of waking, hearing about letting go of penises.

I was vigilant when I awoke, never relaxed. Eyes closed, my attention focused on the movement of the boat—on how it moved in response to the waves. The sea conditions and weather on deck determined the quality of the coming day. Had the weather changed while I slept? Was it a good day out there, or had the weather gone to crap?

To get my spirits up and keep moving forward, I learned to rely on Freddy's spirit. His unflagging energy and upbeat presence helped keep me going. Freddy personified what fishermen call a "good shipmate." He cracked me up, having that unflinching ability to inject humor into any situation, overlooking the negative to focus on the positive. On the worst of days, in the most horrendous weather conditions, I counted on him for a smile, a joke, or a kindhearted ribbing about my work. Freddy's infectious nature helped me survive and keep my emotions from sinking. Sometimes when I listened to him, I noticed that I worked without remembering where I was—this was a welcome and blissful escape from drudgery. Through being around Freddy, I began to look for humor in difficult times.

One time, in the middle of a night that never seemed to end, I observed the crew sitting huddled around the galley table. They all rested with their heads drooped down, faces hidden inside their dark-green oilskins, grasping coffee cups with steam rising into the hoods of their jackets. They were somber, shrouded, and identically dressed, like the monks of a religious order deep in prayer.

Jack drove the boat hard from within the pilothouse, forcing the *Grant* to pound into the large swells, making the boat buck with wild effect inside the fo'c'sle. Cooking pots crashed in storage bins, cans of food thumped at the sides of the lockers, and clothing flung itself around on hooks to hit the ceiling and then whip back to the walls.

The boat's bow flew off a particularly large wave. I watched the crew lift their coffee cups in perfect unison—and they had to, because if they didn't, their coffee would slosh out as the boat dropped. Raising their cups, they looked to me like brethren swinging bells in a bell choir. It looked so funny. I lost it and started laughing—big, deep belly laughs.

Raising his head, Wally barked at me, "What the fuck's wrong with you?" Of all the crew, Wally struggled the most with the power of positive thinking.

"Never mind," I replied. I was too tired to explain.

As the *Grant*'s most subordinate member, I took abuse from the crew. From Wally I took scoldings, chastised like a little brother getting in the way. He had made it clear that I existed at the bottom of the pecking order. I also suspected that Wally resented me for just being there; for being the captain's nephew and having my job handed to me. I felt penalized, not privileged, because being a greenhorn was not a job worth envying. Chris got on me, too, but true to his nature, he was kind. His criticism was not mean spirited. My uncle heckled me in the same way, though he could be rough. Sarcasm flourished in both extremes of Jack's personality—playful jabs when he was happy and hot poker thrusts when he was not.

Jack seemed to pick on Chris more than he did me. He seemed to expect more from Chris, perhaps due to his recent promotion to the status of full-share crewman—and receiving the bigger pay check.

Jack was Jack. He tended to flop things around backwards. He expressed put-downs as if they were compliments, delivering them with his sarcasm. When I had clearly screwed up, he'd say, "That was really intelligent." When he gave out compliments, they sounded like reprimands: "Hey, you're not such a fuck-up after all."

I had grown up with Jack's antics and they didn't bother me that much. I could deal with his abuse, even when I was exhausted.

Freddy and Kaare didn't harass me much. More so than the others, they related to me in a positive fashion, consoling me, making gentle corrections to my errors, and giving me nudges in the right direction.

In spite of all the crap that I took from the crew, in my gut I figured that I was lucky to have these guys as my first crew. They took care of me.

27

END OF
THE LINE

PERMUTATIONS OF FOG, SUN, OCEAN, AND CLOUD SCROLLED across the backdrop of my new life. As part of Jack's strategy over the course of the trip, he chose his fishing spots along the return route back to Kodiak, making each move closer to town. After a move, I might awake to a day of fishing far offshore and out of sight of land. Other times, I enjoyed discovering I was just a mile away from an island's rocky beach, a pristine spectacle of solid earth that made me question whether the boot of an explorer had ever imprinted this wild shoreline. Closer to Kodiak, views of land became more common—snow on high mountains, rock, and treeless tundra.

Our catch of halibut was consistent. Fred said it was "not great, but okay"—a pattern of success good enough to rely on. Like the fishing, the weather had a pattern, too. It was not steady but ran in cycles. Roughly once a week, a low-pressure weather system roared across the Bering Sea, flew over the Alaska Peninsula and dropped into the Gulf of Alaska like a bomb, sending its shock waves across the North Pacific Ocean, tossing our boat and disturbing our lives.

After laboring through a spell of bad weather, I noticed that working on a calmed and glassy ocean was easier than it had been before the storm. Through the experience of each storm, I perceived that I had gotten stronger, both physically and mentally. At times, the emotional work was as difficult as the physical work.

The succession of eighteen-hour workdays had laid down a layer of muscle that wrapped tight around my torso and arms. The strain of work had bent my fingers into a crooked shape, fixed in a claw to hold a knife, chopper, or club. A spider web pattern of grime-filled cracks crisscrossed the calloused skin of my once smooth and soft hands.

Near the end of the trip, I realized that I hadn't looked at myself in a mirror in more than two weeks. Using Freddy's tiny mirror that was hanging next to the stove, I stole a peek at myself and was shocked by the person who stared back; I saw a haggard young man, gaunt and tired. Despite gobbling down giant servings of Freddy's meals, I could see in my face that I had lost a lot of weight. When would I stop wasting away, I wondered.

*   *   *   *

On the final day of fishing, twenty-two days after leaving Kodiak, all routine came to a halt. Upon waking, I sensed energy rushing through the fo'c'sle. The mood of the crew was high and magnified and I joined them in their reveling. Before I had slipped on my layers of clothes, my cheeks hurt from smiling. "Hurrah!" I wanted to shout. "We're goin' back to town." No more baiting. No more setting. We'd haul the gear back and keep it on board this time. I felt relieved knowing that I'd have no gear to bait.

For our last day, Jack had set the gear, weaving the *Grant* through passages separating a group of islands that provided us shelter from the ocean swell. This lessened the rocking of the boat and I became thankful for our stable platform on this day.

The crew yanked buckets and scrub brushes out of lockers. Chris squirted dish soap in the buckets while Wally filled them with water from the deck hose. The buckets erupted in soap suds and a scent of lemon filled the air. It felt good to me to hang onto a tool besides a knife, chopper, scraper, or gaff hook. I put my scrub brush to good use, lathering the boat with passion. Bubbles whipped in the breeze and flew about our heads.

Over the course of the day, we washed the entire fishing boat—stem to stern and port to starboard. Slime, blood, and chunks of bait had covered the boat, crusting in some areas. I had thought that washing my mom's car was a big job. To me it seemed impossible to clean anything as big as the *Grant*.

When the gear started coming aboard, this initiated the biggest job of all. We had to repair the line and hooks—overhaul it—restoring the fishing gear to a like-new condition. Most of the hooks had become dulled at their tips and some portions of the ground line had become worn in the past three weeks. When the rollerman saw a frayed section of line come aboard, he hollered out to us. We cut these pieces out and spliced the line back together, weaving the strands with a fid.

Each of the crew took a turn filing the hooks' tips back to a point. The person who sharpened hooks sat on an overturned bait basket, facing the coiling seat. He grabbed a hook from the coiler as it came aboard, gave the tip of the hook a few strokes with a file, and positioned the hook in the skate before the next one had arrived. My hands cramped up when I tried this job, so I went back to scrubbing the boat clean. My hands wouldn't tolerate hanging on to something as small as a file without cramping. The crew sharpened all of the hooks, one by one—more than 4,000 of them.

The crew behaved like a bunch of partygoers. Freddy was so excited that he shouted most of the day, broadcasting details of what he wanted to do when he got back to town, proclaiming not only that he wanted sex but also what *kind* of sex. Before this fishing trip, I had thought there was only one kind of sex. It relieved me to hear him talk of his own sex life—and not that he was going to hire a prostitute for me.

By tradition, the greenhorn was awarded the raunchiest chore of all— the privilege of cleaning the bottom of the bait table. Over the course of the fishing trip, the top of the bait table had been squirted or scraped on occasion to clean off slime and the big chunks, but nothing whatsoever had disturbed the table's underside. In the past three weeks, slime from the bait had oozed below and turned into a thick, living, rotting mat of goo. As a kid back at home, I was familiar with pond scum. This stuff smelled much, much worse.

I crawled underneath the bait table to scrub at it, and no matter how hard I tried to keep it from splattering, the brush's bristles flung gobs of goop in all directions—on me, and on my face. As I extended my arms upward to scrub, the goo dripped and ran down my arms to soak my shirt. After I finished this awful job, the crew let me to go below to the fo'c'sle, to wash up in the galley sink and to change out of my wet, stinking clothes.

We worked in a frenzy all day, until the moment that Wally raised his hand up high and swept it down to slap the gurdy control handle with a crack to its "stop" position. The sheaves of the gurdy jerked to a halt. Wally reached over the side to lift and swing the final bag and flag on deck.

It was the end. Our fishing was done.

Freddy and Wally raised their hands up to slap high-fives. Jack smiled out the pilothouse window. Chris hollered, "Hurrah!" I joined in with a whoop myself. I felt giddy, knowing that no more fish would come over the side to incarcerate me in the checker. No more coiling. No more chopping. No more baiting endless skates.

The crew seemed pleased with the trip. With its load of fish, the *Grant* felt heavy under my feet. Chris figured we had caught 40,000 pounds of halibut, which would make his share of the catch nearly $3,000. The crew had yet to vote on my pay. I would earn either a portion of a crew's share—a quarter-share or less—or a fixed amount contributed by each man. No matter what they paid me, I knew that I would possess more wealth than I ever had before. The thought of cash in my pocket was intoxicating.

Jack put the boat into gear and revved the engine. The *Grant* sped up to full speed, starting the long run back to town. By radio, Jack had already sold our load of fish to B&B in Kodiak. To arrive on schedule the next day for our delivery, we needed to travel twenty-four hours around the clock. Jack set the *Grant* on a course to bring us abreast of the southwest end of Kodiak Island, cruising along the island's shore.

I took a moment to lean on the rail and gaze and reflect, thinking back to when I had left home and said goodbye to Patti and my family, flown to Kodiak, and sailed out onto the ocean. I thought about setting gear, hauling it back, a stop in Sand Point, coiling, baiting, chopping, and the storms. Excitement overshadowed any emotion that I felt. Had I dwelled on the onslaught of challenges I faced during the trip, I might have cried now, but I couldn't; I was thrilled to be on my way back to Kodiak.

This trip had taken me through hardships along with adventures, making three weeks seem a much longer time. So much so, in fact, that my past and present felt disassociated. I was accustomed to the passage of time feeling linear and smooth. I'd experienced so many things. Time was different.

Freddy had already stated what he wanted to do when he got to town, and now the others piped in. "I'm going to have a martini," said Kaare. Chris said, "I'm going to have a massage." I saw that glint in his eye and tried to guess at what he meant. I had my own wish list: call home and talk to Patti and my family, eat, sleep, and take a shower. But I really wanted to get back in those Kodiak bars, drink some beers, and go see those strippers again.

For me, there was no going back to the past.

28

LOCO-MOTION

From the time . . .

> *the* Grant
> > *left town,*
> > > *the rolling*
> > > *of the boat*
> > > *never stopped.*

> *Never.*
> *Always moving.*
> *It took a while to get used to,*
> *annoying me at first.*

For a long time,

> *I fought.*

It came back.

> *I wanted it to stop,*

> > > *but it wouldn't,*
> > *prodding me*
> > *and poking me.*
> > > *Just when I thought*
> > > > *that*
> > > > > *I could*

                               deal with

                   the rhythm,

        it changed,

          threw me off,

                      bruised me

                and

        beat me up.

               My worst enemy.

            It surrounded me,

        ganged up on me,

and broke me down.

        I got angry . . .

a privilege.

      I was lucky . . .

        to not be seasick.

           I had the energy

             to get angry.

             I resisted.

          No use.

        It wouldn't stop.

      I couldn't win.

    Every second, it moved.

Every second

of every minute,

    every minute of every

        goddamn hour,

           every hour

       of

      every goddamn, fucking

        day.

          The boat moved,

     never

               stopping,

      until

         returning

          to

        Kodiak.

PART FOUR

Layover in Kodiak

29

UNLOADING

THE *GRANT* SLID INTO THE SMOOTH WATER OF CHINIAK BAY and I sank into the deepest sleep of my life. I felt drunk and stoned when Jack woke me up a half an hour later. Emerging from the fo'c'sle into the dawn air, I saw a line of workers standing on B&B's dock awaiting our arrival—one of them, a girl. I halted just seeing her shape.

No sooner had the *Grant* been tied fast than my crewmates launched into action, beginning the process of unloading 40,000 pounds of halibut. Together, we hefted the hatch cover off of the combing, but before we could unload the fish, the remaining ice had to be removed. Almost half of the twenty-thousand pounds of ice blown on board three weeks earlier remained. I donned my oilskins, grabbed a shovel, and jumped into the fishhold with Freddy.

Some ice had melted and frozen again into blocks around the refrigeration pipes that coursed along the "skin" and bulkheads of the hold. Freddy showed me how to swing a steel mallet with short, powerful strokes and then alternately plunge a "devil's fork" into the ice, making it shatter. I felt like a prison convict crushing rocks. Once loosened up, the ice was shoveled into a large bucket that had been lowered into the hold by a cable hanging from a hoist high up on B&B's dock. The crane raised bucket after bucket of ice to be dumped into Kodiak harbor, taking most of an hour to empty the hold. Chris and Wally crawled down into the hold and I climbed out to see that the *Grant* floated in ice, within an island surrounding the boat.

Shoveling ice had pumped up my muscles and I felt crazy with energy, but this didn't have anything to do with working. I came to realize that I wanted to celebrate—to party—regardless that it was morning. I wanted to go to the bars. My duty to work stilled my urge and I accepted that I couldn't go to Tony's right now, but I wanted to go back down in the fishhold and work hard with Chris, Wally, and Freddy. The hatch cover was removed to make a gaping hole in the middle of the deck, providing me a view of them below. They had begun to unload our catch and swung away at the big halibut with gaff hooks, looking like gladiators. It amazed me that they gaffed only fish and not one another's arms or legs. Only a thin layer of oilskin covered their bodies.

To unload the fish, they used the gaffs to snag the fish and slide them onto heavy netting that lay flat on the bottom of the central pen. After the crew had filled the net with fish, they hailed the crane operator, who lowered the crane's hook into the hold. Using loops at the net's corners, the crew clipped the net to the hook. The crew hollered again, and the operator raised the net up nearly fifty feet to the level of the dock. Roughly 1,000 pounds of fish fit in each load.

B&B had a hydraulic-powered guillotine, which spared us the job of chopping off the fish's heads. This allowed us to work faster, unloading fish and cleaning pen boards as the pens were emptied.

B&B's processing crew weighed and recorded the weight of each net as it came across the dock, and Jack kept his own tally to verify them, because B&B would write their check for the delivery based on these figures.

Based on the number of pens filled with fish—nine out of twelve pens—Jack, same as Chris, estimated that we had 40,000 pounds of halibut on board. With each fish averaging around forty pounds, I guessed that we would have about 1,000 fish to handle today.

I scrambled to keep up with the pen boards that flew out of the hold, thrown up on deck by the crew down below. "Goddamn fucking pen boards," I cursed.

I wanted to get off the boat, go to Tony's, drink a beer, and see the dancers. Instead of being down in the bar or in the hold, I was on deck, and I wasn't happy to be there. It began to nag at me to know that the bars were just a quarter mile down the dirt road paralleling the shore. The boards just kept coming and coming out of the hold. I had to wash them all.

The crew scrambled from pen to pen, sometimes sliding on their bellies across piles of fish to get from one place to another. Out of the hatch flowed their chatter, laden with lust for sex and alcohol.

Netload after netload of fish came out of the hold. Unfamiliar with the boat's capacity, I had no concept of the physical volume of our catch. At one point, I thought that the unloading was finished. The hold looked barren except for a pile of slimy ice. Then Freddy and Wally cracked open the aft slaughterhouse, a section deep in the back of the hold that was difficult for me to see from on deck. The three aft-most pens had yet to be unloaded. Would this process never end?

Freddy kept tabs on my progress. When I fell behind and the unwashed boards started piling up, he came out of the hold to help, scaling the vertical walls of the hatch combing like a monkey. By helping me, he helped himself—the sooner we unloaded and cleaned up the boat, the sooner we all could go uptown to Tony's.

After four hours of nonstop work, another net was hoisted out of the hold, and Wally hollered, "That's the last one, boys." The fishhold had been emptied, but we weren't finished. The ice that had been packed in the fishes' pokes and around the perimeter of the hold remained—about two tons of it.

The worker operating the dock crane lowered the bucket back into the hold. I kept washing pen boards while the crew went to work chopping and shoveling ice. They worked at a mad pace, their hair drenched with sweat. Kaare showed me how to dump the buckets and helped me finish washing boards.

A half hour later, the hold had at last been emptied of ice. Kaare dropped buckets and brushes down to the guys, and they started scrubbing the sideboards of the pens and the skin of the inner hull—the walls of the hold. Kaare and I joined them after we finished washing the last pen boards. Thirty minutes of scrubbing with soap and water, and the hold was clean. As the scrubbing was coming to a close, the unloaders—Chris, Fred, and Wally—began departing the fishhold one at a time, leaving Kaare and me to finish the work. They left to go take a long-awaited shower inside the processing plant.

I would be the last to shower. As I finished washing the last of the hold, the others descended B&B's ladder looking fresh faced and clean shaven. They appeared years younger to me without their facial hair. I had decided to let my peach fuzz grow and not shave.

I climbed up the ladder to stand on the dock and became dizzy being on dry land again. Seventeen days had passed since our visit to Sand Point. The swaying sensation continued as I entered the filthy locker room that the processing plant provided for fishing crews. A shower at last, I thought, though I felt uncomfortable opening the valve for the showerhead. For the last three weeks, I had made staying dry my first priority, shunning water and spray. I began to soak myself in water, and something about getting wet on purpose felt wrong.

Inside the stall, my vertigo worsened. I began to feel claustrophobic, feeling that the narrow walls of the shower stall were collapsing onto me. I closed my eyes. The dizziness eased and I began to relax and savor the warm, soft water and the feeling of soap on my body.

Even though I wanted to stay in the shower for a long soak, I couldn't. Jack insisted that my shower be quick. When I emerged from the processing plant with my hair steaming wet, the crew waited for me on board the *Grant*. As I scrambled down the ladder, I heard someone mutter, "About fuckin' time!"

The fo'c'sle reeked of cologne instead of cigarettes, food, and fish. I reached deep into my sea bag to find my town clothes—a sweatshirt and a clean pair of jeans—only to discover that they smelled of pine tar, fish, and diesel. I laid down a thick layer of deodorant to mask the odors.

We threw the *Grant*'s lines from B&B's dock, motored across the harbor, and tied the boat to the transient dock. I was back to where my adventure had started. Of the boats tied here three weeks ago, all had vacated the dock. Others had taken their place. Coming to Kodiak this time, I felt like a transient—a person without a home.

I jumped onto the dock with Chris. It was time to celebrate.

Jack stopped me from running off. He reached over the rail of the *Grant* and stuffed a wad of fifty-dollar bills into my hand. "Thanks, Uncle Jack!" I said. It was my earnings for the trip—$250 dollars. I had a modest bank account at home, but I had never held so much cash in my hands. Earlier in the day, and unbeknownst to me, Jack had polled the crew on what they wanted to pay me. They decided to give me fifty bucks each. I was very pleased with my reward.

Jack had also asked each crewman how much money he needed for the layover, as a draw on his paycheck. They requested anywhere from $200 to $500, except for Freddy. He wanted $1,000.

Together, Chris and I sprinted uptown.

30

HASHISH, FLIPS,
AND A WHORE

I ENTERED THE BAR, HOPING THE BARTENDER WOULD LOOK the other way, and settled in at a table with Chris. Next to the stage, the jukebox rattled and thumped with a rock song. I recognized some of the dancers from before and saw that nothing had changed. It was dark and sleazy, and it reeked of beer and cigarettes.

Several times, I ducked outside to a pay phone in the breezeway and tried to call Patti. I struck out. She wasn't at home or at work. I felt more anxious with each failure to find her and talk to her.

Freddy joined our table and we started to party. Chris and Freddy bought round after round of drinks, keeping me supplied with beer and helping me conserve my money. About the time I started to feel tipsy, Chris turned to me, telling me that it was my turn to buy a round. I put a fifty on the table. It pained me, when I related the cost of the drinks to the amount of work that it took to earn fifty dollars.

I spent several hours with Chris and Freddy and I was getting drunk. Chris hauled me next door to eat dinner at Solly's Bar and Grill. The food sobered me up just enough to return to Tony's.

One advantage of partying with Freddy was his frequent purchase of table dances. A table dance cost five dollars and lasted as long as a song played on the jukebox. Table dances were an up-close-and-personal talent show of buck-naked charm.

Chris and I sat at Freddy's table, with me hardly a foot away from his dancers. I blushed, hearing Freddy coo at one dancer in a soft purr. The dancer's expression was unfazed when she brushed her legs against his. I gazed over the girl's curves—tawny skin stretched over a snaking body. This was not a glossy fold-out from a *Playboy* magazine. This was a living, breathing woman in front of me—full breasted and bare assed.

As an adolescent watching TV at home, I'd seen animalistic displays, watching shows like *Wild Kingdom*, squirming in my seat when a male lion in the African savanna took a whiff of the haunches of a lioness, lurched on her back, and started humping away. Here I was, a young man, behaving like I was in a nature show—with the narrator droning, "Observe the young males in the company of females. . . ." Feeling guilty, I was like a kid who licks his finger and keeps on sticking it into the sugar bowl—oh, but it tasted so good.

After one more hour of drinking and buying another round of drinks, I couldn't take any more. I couldn't sit still. I had to get out of Tony's. I had to move. Chris and I left the bar, abandoning Freddy, who continued to engross himself with the strippers.

Leaving Tony's, Chris and I tried to get a drink at the Breaker's Bar. The patrons of this bar were hardened. They looked tough and they acted tough. When I failed to present I.D. to the bartender, I felt relieved when he ordered me to leave the premises.

Chris hailed a cab—a rusted-out station wagon, its interior caked with dust—telling the driver, "Take us to the Beach." Like the cab from the airport three weeks earlier, this cab reeked of stale beer and cigarettes.

As the cab pulled out, Chris lit up a joint and offered me a hit. The cab filled with pot smoke, but to my surprise the driver didn't say a word.

The Beachcomber Bar—the "Beach" to all who knew her—was a ship. It had been hauled up onto the beach as a post-earthquake, emergency watering hole, and it sat high and dry in the middle of a gravel parking lot. From outside the ship, I could hear the muffled sounds of a rock-and-roll band banging out a Led Zeppelin tune. I had a hunch I'd like the Beach.

Chris and I hiked up a gangplank and boarded the ship. I ducked to get through a low doorway to find the bar in the main cabin, weaving down a narrow passageway lined with a row of portholes. Entering the dim light of the bar, I saw the rock band playing on the other side of the room. A few couples danced. Chris and I ordered beers.

As we sat, enjoying the music and watching the dancers, a woman entered the room, breezing in with a show of charisma. Her platinum-blonde hair looked like cotton candy and was done up into a bouffant hairdo that towered over her head. She held her face in a way to express regality.

Chris nudged me in the ribs, put his hand over his mouth, and whispered in my ear, "That's Dee. She's the queen bee . . . the madam of the whorehouse."

Silent, I mouthed the words, "Holy shit." My body tightened as I watched the bartender and patrons in the bar acknowledge her, exchanging nods and knowing smiles. Fearing that Chris might get her attention to set me up with a hooker, I hunched over and became motionless, wishing that I were invisible. I guessed that her appearance here at the Beach was meant to show that her business was open and ready for customers. In less than a minute, Dee disappeared out of the bar. I was safe.

Well after midnight, I started to nod off from exhaustion and too much to drink. I resigned myself to going back to the boat. I couldn't afford this lifestyle any longer. Drinking had tapped my wallet. I caught a cab back to the transient dock.

I wouldn't see Freddy again for several days. The next morning, he left Kodiak, flying to Anchorage for more serious partying. Chris said Anchorage was better suited for satisfying Freddy's needs. It had an easy supply of heroin and a greater selection of strip clubs and whorehouses. Chris had decided to stay behind.

* * * *

That night, lying in my bunk on the *Grant*, I wrote to Patti before I passed out. I scribbled with scarcely readable handwriting.

July 29—2:00 AM
Kodiak! WOW
    First of all, I must admit that I have gotten bounced out of 2 bars.
    #2—I am droned (dr-drunk, oned- stoned). Just got back from the
Beachcomber Bar (far out place). I was watching all the people dancing with
their girlfriends and was thinking of you. It is really lonesome up here. My

fellow shipmates have told me to warn you to stay away from me for at least one week.

Anyhow, I love you (and that's the truth, pphhhh).

*   *   *   *

The next morning I awoke with a headache. My spirits rose upon realizing that the boat was still docked in Kodiak. The fo'c'sle was quiet, except for the stove fan droning away with the kind of noise that, if you turned it off, it'd be so quiet that you'd want to turn it back on.

I got up moving slowly, trying not to disturb Chris and Wally in their bunks, and got some cereal for my breakfast. I poured some ultra-pasteurized Real-Fresh in my bowl and cursed at the carton. I promised myself to buy a carton of fresh milk when I went uptown.

I had little to achieve this day—call home, take my laundry in to be cleaned, look for a card to mail to Patti, and buy milk. I'd grab some on the way back to the boat. I brushed my teeth and rinsed my toothbrush, moving the pump handle, trying to be quiet.

I climbed the ladder to the deck and a ray of sunlight shot into my eyes, making my headache flare with pain. I retreated into the fo'c'sle for my sunglasses. Finally, with my sea bag hanging over my shoulder, I wobbled up the dock and shuffled over dusty streets and sidewalks, making my way for B&B. My head felt heavy, but my tennis shoes made my feet feel light. They were like bedroom slippers compared to my rubber boots.

I recognized the Breaker's Bar in passing and came to St. Paul's Square, a small square with benches, surrounded by bars and a drugstore. I was shocked to see bodies on the benches, lying motionless and askew. Without exception, all were Native Alaskans—drunk and unconscious. I shuddered, thinking that they must have been freezing just lying there. A feeling of helplessness forced me to leave, yet the sadness of the scene stayed with me as I walked on.

Reaching B&B, I climbed the stairs to the office. I asked the lady behind the counter about getting my clothes washed—a service that B&B provided to fishermen for a fee. Following her orders, I put my bag in the corner of the reception area and left, happy to be freed of the burden of doing laundry. I didn't want to work today.

I rounded the corner toward the processing plant adjacent to B&B.

The sign on the building nearest the dirt road said "King Crab, Inc."
I thought, "What the heck," and though the smell of seafood became
more rank as I continued, I followed a passageway that led out to King
Crab's dock. I knew that crabs were caught in the winter, but I hoped to
find one of the giant crustaceans.

Two steel boats were tied to the dock. They were painted in bright colors and looked brand new. Close to 90 feet long, these boats were much
larger than the *Grant*. They looked different, too. Their pilothouses were
built onto the forward part of the vessel, leaving the stern deck as a large
work area. One of the boats had shrimp piled into an enormous pink-
colored mound that covered its entire deck—millions and millions of
shrimp, the smallest of crustaceans. The boat rode so low in the water
that I expected it to sink at any moment. The hold must have been full of
shrimp, too, I guessed. My mouth hung open at the sight.

A crewman on the shrimper stood on top of the mound in his oilskin
pants, his legs sunk into shrimp up over his knees. One hand rested on
a shovel stuck into the pile. His shoulders were broader than his dingy
T-shirt with its sleeves torn off, and his muscular forearms and biceps
bulged. He stood with his head cocked over his shoulder to look at the
dock, his chin raised and jutting out, in a pose that reminded me of
Soviet poster art. Heroic.

I was fascinated by the shrimp boats. Compared to the 45-year-old
Grant, these boats seemed like dreamboats. I found them attractive, the
same way I felt about V-8 hot rods. I stopped to admire the boats and
the load of shrimp, and then turned to see if I could get into King Crab's
processing plant.

When I entered the plant, a worker stopped me. Raising his voice over
the noise of machinery, he hollered, "Ya gotta wear a hair net in here."
He held out a tangle of fuzz for me to put over my head, but I shied away
from putting it on and raised my hand to say no.

Stepping outside, I peeked in through the doorway. I had a good view
inside of a large machine with a series of angled rollers lying parallel
to one another. Shrimp entering at the top got pinched between rollers
that peeled off the shrimp's shell. A pink morsel of shrimp meat stayed
on top of the rolling peelers, and the shell fell below. The peeled meat
passed down the rollers till it fell onto a conveyor belt that sent it out of
my sight.

This machine reminded me of the wringer that my grandmother had once used with her laundry. She'd crank on the machine's handle to make two rollers turn, and she'd feed an item of clothing into the narrow space between the rollers, squeezing water out of the fabric as it passed through the gap.

I stood there, mesmerized by the hum of the machines. When I came back to my senses I decided to press on with my errands.

The grocery store was my next destination. After walking along the waterfront on the dirt road that ran into town, I came to the town's square—a parking lot surrounded by several bars and shops.

A young man was clomping along the sidewalk in rubber boots, and I asked him for directions to the grocery store. He had a disheveled look—grubby clothes, mussed up hair, and a week's worth of stubble for a beard. Not much older than myself, I guessed that he was a fisherman, too.

"Kraft's? It's just up ahead, through the breezeway."

It was the same breezeway that separated Tony's and Solly's, reeking of beer, cigarettes, and bile. I stepped over a puddle of puke and weaved to avoid steeping on the shards of a broken beer bottle. Rounding the next corner, I was surprised to see that Kraft's was a tidy and modern supermarket.

I growled when I saw the price of a carton of milk. "Two dollars?! Shit!" I reached in my pocket and pulled out my wad of cash. Counting it, I found that I'd spent almost fifty dollars the night before. I had two more nights to go in Kodiak. To hell with milk. I'd wait until Freddy came back for the next trip. He'd buy fresh milk. It was then that I realized I was making plans . . . for the next trip.

When I had left home for Alaska, my plan was to work until September when school started, not knowing that this job would be so difficult and brutal. During the toughest days of fishing, I swore many times that I'd never allow myself to work on the ocean again. Now, however, after only one day in Kodiak, I was already retracting this promise.

The possibility of making a few hundred more dollars fueled my urge to continue fishing. I knew I could do better on the next trip—and make more money. Aside from the opportunity for more pay, I liked some of the side benefits of this job. I thrived on being away from home and I relished my independence. That's all there was to it. I'd finish what I had set out to do. I'd make the second trip.

To find a card for Patti, I had to pass once more through St. Paul's

Square on the way to the drugstore. The residents of the square were awake and ready to get drunk again, grousing to one another, squawking like city pigeons, oblivious of the world around them.

In the drugstore, I searched for a card for Patti from the rack. A romantic message with swirly writing? No. It wasn't my style. I felt depressed thinking about how much I missed Patti, so I settled on the one with the joke inside that made me laugh.

One more stop—the phone call.

I had noticed a phone booth on my way into town. I found it and discovered that the phone had been vandalized, its handset ripped out. The next pay phone I found was similarly mutilated, and I wondered why. Had the vandal been ditched by his girl in a phone call? Had he received some other really bad news from back home?

Outside the post office, I found a phone with a dial tone. I paused before dialing, remembering that two time zones divided me from Seattle. Noontime in Kodiak was 2 p.m. in Seattle. I called Patti's house first and her mom answered. Shy with her, I avoided making conversation and just asked her where I could find Patti. She didn't know.

I tried another number, calling Patti at the ice cream store—the same job I'd left. No luck. Striking out, I decided to try calling my house. My supply of quarters was dwindling and I dialed the operator to make a collect call, figuring that my mom would be home from work by now. Mom answered after a couple of rings and she sounded relieved, saying that she had been waiting to hear from me for some time, several days in fact.

The call was awkward, long-distance at its worst. Each time I spoke to my mom, there was a delay, and then I heard my own words echoing back through the ear piece. After that, I'd hear my mom answer, followed by the same echoing delay. We stumbled through our conversation.

I told her about the fishing trip, and that at times during the trip I had been certain that this was going to be my first and last. "I was sure I'd never go out again . . . but I've changed my mind." I paused, but she didn't reply. "I've decided to make another trip."

She stayed silent.

"Fishing is horrible during the storms," I said, "but when the weather's nice, it's almost fun."

She talked again when I asked her about my brother, "He's with Nick and Pete right now . . . at the Lake."

All at once, loneliness crushed me. I missed home. I missed playing on the Lake. I missed Mom, Patti, and my brother. I could guess that Jon was waterskiing with our friends. Kodiak was a world away from home.

When it came time to finish our call, Mom spoke softly, clipping the words when she closed. "Take care of yourself. Call me when you can."

I hung up and tried calling Patti one more time. No luck. I felt miserable now and trudged back to the *Grant*, dragging my feet through the dust. It lifted my spirits some to find Chris in the fo'c'sle, reading a magazine in his bunk. I was glad to see him.

*　*　*　*

My second night in Kodiak resembled the first. I had made it back to the *Grant* in the early hours of dawn.

July 30—2:30 AM
I am in a little bit better shape than last night. We had a so-so trip and I got $250 bucks.

Called home today and I figure the only way I can call you is person-to-person (seeing as you are all over the place). I hope you aren't working when I call.

Kodiak gets boring after awhile because I am alone (no fun). Went on a drive around the island and found a WWII bunker.

Chris and I spent most of the next three days together. We slept on board the *Grant*, ate our meals together, and paired up for the beginning of our nighttime adventures into town. Though I felt isolated at the transient dock, away from the activity of town, I avoided the bars as much as I could. Bar-hopping had depleted my wallet and the ensuing hangovers had hammered my well-being. With Freddy gone, Chris and I rummaged through the food lockers and survived on canned food left over from the fishing trip.

Toward the end of the layover, weather moved in and it started raining.

For our last night in Kodiak, Chris heated up some soup for a snack. Wally was gone. He had splurged by renting a room at the K.I.—the Kodiak Inn. I climbed into my bunk to read my book, *The Godfather*.

"You want some soup?"

"Sure," I replied.

"I'm not going uptown tonight. You wanna play some cards?"

"Yeah, I'll stay, too. I can't believe how much money I've spent."

He chuckled, "Yeah. Me, too." He nodded. "Hey, I'm going to buy some wine and make flips. You want some?"

"A flip? What's that?"

"It's red wine mixed with Seven-Up."

"Never had it, but it sounds good."

"Okay, I'll buy enough for you, too."

I nodded off to sleep and awoke to find Chris sitting at the table reading a book. Several hours must have passed. It was dark outside. Chris had a jug of wine and a six-pack of pop in front of him. My attention became keener when I noticed the smell of pot in the air.

Chris mimicked Jack's voice, asking me, "Hey, Sleeping Beauty. Ya want some?"

Chris was stoned.

I climbed out of my bunk.

"Here," he said, reaching out with the bottle. "I splurged and got the good stuff. It's Gallo Burgundy."

I smiled and raised my eyebrows, implying, "What else have you got?"

When he smiled back, his eyes squinted shut behind his glasses.

"Oh, yeah. Before he left, Freddy gave me a little treat." Pinched between his fingers, he had a small wad of foil and handed it over to me. "It's hash . . . Turkish hashish."

"Whoa!" I said, in amazement. I unfolded the foil. Inside, I saw something herbaceous that reminded me of a rabbit turd with squared-off edges.

Pot was within the reach of my circle of friends at home. Hashish was not. It was mystical and exotic. Before that night, I had never smoked hash nor even seen it. I'd only seen pictures in health class. Instead of smelling like a grass fire the way pot did, it smelled more like incense. Chris lit a candle and told me to turn off the fo'c'sle lights.

Absent the harsh illumination of the bare bulbs, the fo'c'sle seemed like a hideout, a place safe from Kodiak. Chris and I had a great time that night, enjoying playing cards, drinking, smoking, and laughing.

I finished the night with a burst of writing, adding the following entries to my letter to Patti:

August 1—Well, Kodiak wore me out and all there is to do here is drink, which can get pretty boring. I am more tired here than out fishing.

Ask Weasel, if he has cracked up a car lately and tell him that my cables are whipping his cobwebs.

"Weasel" was one of my best friends at home. We teased each other about the fuzzy hair of adulthood showing up on our faces, referring to our own whiskers as growing thick like cables and the other's as being wispy like "cobwebs."

I might get $400 if I am lucky. Have fun at B&R. The only thing the same about B&R and the boat is that Jack likes his bait a certain size (and not 2½ ounces, thank God!).

Tell Ferd—keep the Mafia in order. By the way, I have read 2 Mafia books on the boat. That party is going to be a good one.

Below that remark I drew a crude sketch of four kegs of beer. "Ferd" was another of my friends. The "Meridian Mafia" was what we called our group of neighborhood kids.

Don't do anything I wouldn't do (a nice cliché, but at least I am giving you lots of leeway). That address is a hint for you to write. I hope you have a nice tan (I am fading badly). Weight-wise, I am fading (165 in the winter and 148 now). Better watch it, or else old "FATSO" yourself, will outweigh me (HA!).

Please write—soon (before Aug. 10)

Love, Dean

The bit about Patti's weight was a joke; her body was trim. I would have died had I shriveled down to her size.

In the letter to my family, I wrote:

August 1—Just iced and got bait. We are fogged in and we cannot get Freddy (the cook) in from Anchorage. Will leave tomorrow. I really weigh 148 (lb.). I must be working a little of my love handles and my seat pads off, or something. When I called you, I was still used to a different time zone called "fishing time zone," which is working from the afternoon, well into the next day. I am beat. Can't wait for my vacation. Have read the book "A Clockwork

Orange" (gory), and I am reading "The Godfather" (gory, also). I will be reading "I Bury My Heart at Wounded Knee" as soon as I can get some "rest" out fishing. Anyhow, I will NEVER regret quitting my job at that "other place."

On the margin of the letter, I drew a picture of my boss at the ice cream store straining to pick up an imaginary object that I labeled as weighing one pound.

It's not so bad up here. Anyhow, I still miss everybody (even Jon, and how is that for weird. Wow.).

To Mom: Believe it, or not, I am still your son, and can't wait to get home. Have you got the new car? The address is for writing and giving me any specific instructions on coming home. (I am planning a secret return).

To Dad: They didn't get me any "crotch-rot" material. (Thank God.) They were really pressuring me.

"Crotch-rot material" refers to prostitutes, venereal infections, and the seeping sores I'd seen in the slideshows of junior-high health class.
More for Dad:

Watch Jon for me. We have had beautiful weather for a while now and this trip I might get some rotten weather. I will be doing everything in the way of regular work, because Wally will teach me some stuff on the "gurdy-roller." I have speeded up on baiting and dressing. Wally reminds me of you, the way he kids me about being slow and when I slip, or something.

Well, even though I miss you all, I am ready to go out again.

Love, Dean

PART FIVE

The Final Trip

31

SANDMAN REEFS

JACK CROSSED THE DECK, FUMING MAD, AND CURSING AND using Freddy's name while he stormed by me. Then he cursed at the fog that had kept Freddy's plane from landing in Kodiak.

That day, we outfitted the boat for the upcoming trip without Freddy, taking on ice, fuel, and bait. For our bait, Jack had bought from B&B about 10,000 pounds of frozen herring, octopus, and some pink salmon—"humpies." While we loaded the salmon into the hold, I noticed that it had a faint odor of eggs. It bothered me right away, making my stomach feel queasy. Jack smelled it, too.

"I think it'll work for us," he grumbled. "They gave us a good deal on it."

By late morning the next day, the clouds had broken and Freddy's plane flew over the harbor. He arrived at the *Grant*, looking more exhausted than after the fishing trip. Jack shot him scathing looks when he stepped on board.

I ran up to Kraft's with Freddy, helping him shop for food to last us three weeks. Freddy flew down the aisles of the store in a whirlwind, filling a half-dozen carts. As soon as the grub was loaded onto the *Grant*, Jack ordered us to throw off the lines and we cruised out of Kodiak's harbor.

Once again, we traveled to the west. The *Grant* bucked into heavy seas for two days, forging through waves—up and down, up and down—endlessly. The essence of rotting salmon wafted into the fo'c'sle, torturing me in my bunk and turning my stomach. I felt horrible and wished that I could stop the *Grant* from moving. A throbbing pain chiseled at the inside of my head, like a hangover headache, but I hadn't done any drinking before we left town. I ended up blaming my condition on the spoiled bait, *not* seasickness, citing as proof the fact that I didn't throw up.

All of us were battered by the weather. The rest of the crew looked as haggard as I felt. Twice a day, we emerged from our bunks for food, crawling from crevices like vermin to paw and nibble at our food, then skulking back to our nooks in the walls.

Freddy prepared our meals under the most hideous of conditions. He cooked in a cramped kitchen that continually sprang up and down as though it bounced on a trampoline. In reaction to the worst waves, pots and pans clanged, food spilled, and Freddy cursed. In nice weather, the fo'c'sle was a tough place to cook; in bad weather, it was a hellhole.

Twice a day, I served my two hours of wheelwatch duty, nauseated or not. I wedged myself into the pilothouse seat, stared at the compass, and watched the bow bounce up and down, and up and down. After my wheelwatch, I dove straight into my bunk. When I wasn't sleeping or *trying* to sleep, I spent my time like the others, reading books. Reading was a good way to pass time, but my headache plagued me, forcing me to put my book down. I wrapped my pillow tight around my head, trying to mute noises that were now intolerable—the reverberation of the propeller when it cavitated or the cacophony of seawater thrashing across the deck.

August 5—Well, with the combination of a rolling boat, rotten salmon bait, I got the biggest and longest stomach ache (2 days) in my life. No actual sickness, though.

On the second day of traveling, Jack called me for a wheelwatch. Rising from my bunk, I noticed that the boat barely rocked at all and the throb of my headache had lessened.

Jack awaited me in the wheelhouse with his head stuck out the window. "How ya doin'," he asked.

I sighed, "Much better."

"Glad to hear it. You looked like crap yesterday. Calm weather does one good."

I nodded back.

The *Grant* was traveling through a pass between rugged shores.

"Where are we, Jack?" I asked.

He pointed behind the boat—to the starboard side.

"That's Cape Kupreanof, there. It's on the Alaska Peninsula . . . the mainland. The Shumagin Islands are off to the port side." He pointed to the islands sheltering the waterway, protecting us from the weather and the ocean swell of the North Pacific. The calm water was restoring my energy, along with my curiosity.

"Where are we headed?" I asked.

He grinned, saying, "That way," pointing ahead to the bow.

"Give me a break," I said, laughing. "What's our destination?"

He stared out the window now, his expression grim, replying, "Sandman Reefs."

This didn't make any sense to me. Boats don't head for reefs.

Jack could tell from my silence that I was puzzled.

"Here, I'll show you," he said.

Jack ducked through the door into his stateroom, which was so small there wasn't enough space inside for two to stand. I stopped outside the door and looked over his shoulder. He reached to the ceiling and unclipped the chart table. The chart table hinged down to become angled like a drafting table, with rails that corralled his nautical charts. Two strips of wood lay tight across the top of the charts to hold them fast.

Jack freed one of the wood strips to sort through the stack. The top charts curled around his arm. He scanned numbers printed in the corners of the charts, selecting *"8802"* from the middle of the stack: *Alaska Peninsula and Aleutian Islands to Seguam Pass*. He jerked *8802* out of the pile, swung it behind his back, then up and around in front of him to rest flat on top. He reminded me of a bullfighter sweeping his cape— *"Ole!"*

Clamping it down with the wood strip, he slid his index finger westward across the chart, starting at our current position and stopping at a point north of an island called S A N A K.

Throughout my childhood, my uncles' fishing charts had fascinated me. By studying navigation books and asking questions, I had learned the basics of reading the charts, the numbers and a variety of symbols showing depth of water, hazards, landmarks, and lights and aids for navigation.

Once I had called a chart a "map." One of my uncles—I can't remember which one—corrected me, "No, no, no. It's a chart. Maps are for the land, and charts are for the sea."

Looking where Jack pointed, I saw a pattern of marks indicating foul ground and rocky hazards, encompassing an area more than twenty miles in diameter—a nightmare of navigation. It looked like the symbols had been sprayed onto the paper with a shotgun blast. The center of this chaotic pattern of asterisks and crosses was labeled SANDMAN REEFS. I didn't know then the difference between an asterisk and a cross—marks for submerged and exposed rocks—but I knew that they both identified rocks. Blank space surrounded a good portion of the reef, meaning that this area had never been charted.

"That's it!" Jack sighed. In his eyes, I glimpsed something akin to the glint of an old-time pirate hunting for treasure.

Jack guessed it was unlikely that other boats had fished in this area, considering the complexity of the reef, leaving it as rich, virgin ground. He chose to fish here, he said, because many species of fish were known to congregate around rocky areas, including the large halibut that ate them.

"Wow. Looks nasty," I said. "How do we get in there?"

"We don't. Well, at least not right away."

"Why not?"

"I can't trust these charts, especially this one. It's too big. There's not enough detail . . . lots of places are uncharted. Funny thing, even when rocks are shown on charts, you still can't trust 'em. Some charts indicate rocks that don't exist."

I turned to look out the windows, scanning out over the bow to make sure that we held a safe course, with no uncharted rocks ahead of us.

Jack continued. "First, we're going to just poke around and check out the perimeter of the reef. I did that a couple years ago. Fishing was pretty good and I'd like to try it again. I just listened to Peggy, and she gave us a nice forecast. If the forecast holds true, we just might be able to go inside the reef and fish in the uncharted area." He turned to me and said, "That's something I haven't done before."

I didn't know whether to be excited or scared.

"Your watch is two hours long, starting at ten," Jack said.

He disappeared into his stateroom, shutting the door. I sat alone in the pilothouse, savoring the stillness of the boat and the sudden return of my appetite.

The *Grant* traveled on flat water now. In the space of minutes, Chris climbed up out of the fo'c'sle and took a piss over the rail. Inside the companionway, I saw Wally's head stick up. Freddy came on deck and popped through the pilothouse door to find out what was going on.

I told him, "We're going to Sandman Reefs." It felt exciting to be the herald of Jack's fishing adventure. Freddy just furrowed his brow and said nothing, looking immediately concerned.

After Freddy made breakfast, we suited up in our oilskins, except for Jack, who was on watch in the wheelhouse while we traveled. The beat of the chopping block returned to the deck and we started to bait gear. The layover had officially come to an end. It was time to go back to work.

A breeze whisked across the deck, taking with it the stink of the spoiled bait. It turned out that only the belly of the salmon was rotten, because the guts of the fish hadn't been removed before they were frozen. The acids in the gut had burnt the bellies. We trimmed the belly flaps off the fish before chopping the salmon into bait. We baited, took a coffee break, baited some more, broke for lunch, and then more baiting. In five hours, the gear was baited and ready to go. In no time, I'd become spoiled sailing on calm water. I knew I was going to miss working on a steady deck.

The *Grant* cruised out from behind the shelter of the islands and began to head offshore. I groaned when the boat responded to the first ocean swell and began rolling again. I went down below to grab a nap before we set out the gear. I had three weeks of hard labor ahead of me, and little sleep.

As I nodded off, I puzzled over the name of the reefs—"Sandman." Where did that come from? I knew the tale of the sandman putting children to sleep. Perhaps someone had fallen asleep here at Sandman Reefs. Was the name associated with some kind of disaster?

* * * *

I awoke to sun and a blue sea. Emerging from the fo'c'sle, I looked over the horizon to search for the reefs to find nothing but water. Jack stuck his head out of the pilothouse window. I yelled up, "Where's the reef? I don't see it."

"Oh yeah, it's here," he replied, an edge to his voice. "I can see several rocks."

I stepped up to the level of the poop deck and followed the direction of Jack's gaze. First one, then two rocks appeared, very distant on the horizon, looking like scabs. As I watched, a swell larger than the rest broke over the top of one of the rocks. Like a geyser, a crest of whitewater erupted from the blue ocean and fell back over the rock. The explosion of spray looked as if a stick of dynamite had been ignited underwater.

"Holy shit!" I murmured. If a reef could make a wave do that, I shuddered to think what it could do to a boat.

For the entire two hours of setting gear out the stern, Jack never paused or rested, whirling around time and time again to look at the fathometer screen, checking how much water we had under the boat. He looked this way and that, grabbing the binoculars to scope out a new rock we had come upon. After we finished setting gear, Jack looked so wound up that I wondered if he'd be able to relax before we started fishing. I hoped he'd get a few hours' sleep while the gear soaked.

*   *   *   *

On the first day of fishing we caught more fish than on any day of the previous trip—5,000 pounds. In spite of the excellent fishing, the crew grumbled and stomped around deck. They spoke few words, only those necessary to operate the boat—calling "SKATE OF GEAR!" or "GIVE ME A HAND"—and little more. I missed the typical chatter of the deck. Without it, fishing just wasn't as much fun. At the same time, I was impressed that we could accomplish so much work in silence. I guessed that they felt like me, lethargic, pummeled by the rough weather on the trip out. My strength had been sapped.

By the second day, I came to realize that the rules had changed, though no one said as much. The whole crew was on my ass. While we hauled gear on the first day, Chris had barked, "Hey, snap out of it, goddammit." The harsh delivery of the "goddammit" caught me off guard. The moment I started to space out, someone jumped on me, whether I was examining the contents of a halibut's stomach or gazing over the ocean, marveling at an albatross that swept through the air in wide arcs. Everyone was watching me.

"Hey, Flash. Quit fuckin' the dog," Wally yelled.

Something about the way Wally insulted me started to get on my nerves. He kept prodding me to move faster. More and more, he was becoming my taskmaster. This infuriated me. I dug in my heels and resisted, becoming angrier. At times, I wanted to lash out, tell him to go fuck himself. On the previous trip, I had been able to deflect Wally's and the crew's needling and not let them get to me. What I couldn't yet see was that I was working faster and harder now—more work and more results.

Freddy had taken the role of technical advisor. He showed me ways to work more efficiently, to simplify the extra work that Wally gave me. In combination, Freddy and Wally were "good cop/bad cop." They forced me to keep my attention on board the boat, keeping my mind *right here* and *right now*. Whether I liked it or not, I could no longer daydream.

On the first trip, when the crew had left me alone to work, my pace had slowed until I traveled the path of least resistance, adopting a tortoise pace. This trip, if I didn't give it the extra mile on my own, they were going to prod me until I did. To become a complete fisherman, I needed to stretch to reach the next level and learn what it was like to hustle in my work. Still, I had a long way to go in terms of building strength and endurance.

In terms of emotional endurance, I felt overwhelmed just thinking about the duration of the trip that lay ahead. Three weeks at sea seemed an eternity, too long and too far away from home.

By the third day of the trip I'd caught a cold and was drained of energy. Like all ocean fishermen, I worked on my sick days.

August 8—Now I have a good "cold." Good fishing: 1st day—5,000 lb.; 2nd—4,800 lb.; 3rd—5,000 lb. I hope we can keep up the good work. Coiling is my goal right now. I coiled a half-skate and just when I caught on, it got kinky. I think I can do it now, though.

*   *   *   *

A boat on the high seas is isolated in many ways, which makes it virus free, that is, until someone brings a virus on board. Then, in effect, the entire crew is trapped on a boat with a Typhoid Mary. The crew hassled me, telling me that if they caught my cold, I would be on their shit list.

I noticed that Jack's role on deck had changed and he was acting in some ways like a greenhorn. He had stopped ribbing me, which was extraordinary. He had teased me and given me a hard time since I could walk. Now he was quiet on deck. He started the trip conservatively by deciding to keep some of our skates on deck, not to set them all out, skirting the fringe of Sandman Reefs and using only a portion of our gear. A few days later he seemed to have become better oriented to the surroundings. He was less tentative, expressing more confidence. "I think I know where we are now," he announced. Though Sandman Reefs appeared disorganized on the nautical chart, Jack had begun to ascertain a pattern, a degree of geological order to the reefs. He was gaining faith in his ability to estimate the *Grant's* position on the chart and to guess where the next reef or rock would emerge. Having taken steps to assure himself of our plotted position, he now felt ready to go within the reef's perimeter. It was time to venture into the unknown, time to accept the danger of the reef, and time to take some chances. I had confidence in him. With all the rocks around us, I'd have gone crazy if I hadn't trusted Jack.

Starting with our first day of fishing inside the reef, Jack's gamble paid off—our catch of halibut jumped to a new high. In the days that followed, our daily catch doubled those of the previous trip. We were averaging more than 4,500 pounds each day.

Fishing inside the reef presented other hazards. The rocks that jutted through the surface of the ocean represented the tip of the iceberg. The seafloor was lined with jagged rocks. Each string we hauled got snagged, or hung-up, on the ocean bottom, so much so that Jack relinquished the role of dealing with the hang-ups to the rollerman currently hauling gear.

Sometimes the rollerman brought the gear tight and was able to wrench it free. Other times, the gear parted. When we hauled from the other end of the broken string, we feared that the gear would snag and break again, resulting in a loss of the middle section.

Our stock of gear dwindled as we lost portions of skates and strings, but Jack said that our good fishing within the reefs far outweighed our cost of losing gear.

August 9—4,500 lb. I might be home pretty early if this keeps up!

We enjoyed the heavy fishing and the extra work that it entailed. Our continued catch clearly indicated that a large concentration of fish occupied the reef area. If this trend kept up, it was a sure thing that we would fill the boat, a 70,000-pound load that would translate into a fat paycheck for Jack and the crew. Considering that I had contributed more toward this trip, I had hopes of doubling the $250 I had earned on my first trip.

Based on our current catch rate, I forecast that not only would we fill the boat with fish but we would also finish the trip early, before three-week's time. This meant that I might get home before summer vacation ended, get more rest, and enjoy a few days on the Lake before going back to school.

Jack's mood swung back and forth, from relaxed to tense. He was constantly on the alert, spending more and more time looking out over the water, scanning the horizon for rocks jutting through the ocean's surface. Whether working at the roller, tending to hooks as they came aboard, or coiling a skate, he lifted his head to look out over the waves, watching for an uncharted reef. There was more at stake here than our safety. We had a valuable load of fish on board.

Every couple of days, we had caught enough fish to fill an additional pen in the hold. To make more room, the crew sent me down below to shovel ice out of the way. It began to go hand in hand, the more fish we caught, the more ice I shoveled. Already the boat felt heavier, more burdened.

In the second week of the trip, the weather stirred, but it didn't slow down our fishing. Instead of the wet, dreary low-pressure weather systems that plagued our first trip, a strong high-pressure system sat on top of us, providing sunny weather with strong winds. Despite the sun, we had to don raincoats to stay dry from waves sloshing over the rails.

When foul weather came, the noise of the wind and waves took over the deck and the crew's talk quieted. Of the words spoken, the wind shredded the holler of "SKATE O' GEAR" to a phrase recognizable only by the rhythm and punctuation of its syllables. I was never sure of the emotions of the other crew members because we rarely spoke. The waves had silenced us.

The crew's hazing of me had eased some. I worked harder and faster now, gaining in speed and expertise. They had less to complain about. When the crew hailed out, "Hey, Flash. We need more bait," I knew just what to do. I jumped to work, splitting up codfish or salmon and chopping it. I did my job, not needing someone looking over my shoulder at every turn.

On one sunny and calm day, I baked a cake for the crew—in part for being thankful with the slack that they had cut me. The cake came out of the oven and I brought it on deck, warm and dripping with frosting. Jack and the crew devoured it. Wally smiled, with chocolate highlights hanging from the corners of his moustache. For the remainder of the trip, Jack called on me to bake cakes any time the weather lay down.

For two straight weeks now, we'd enjoyed good fishing, forcing us to adapt. We were shorthanded, having stationed a man in the checker to clean fish most of the day. Mealtimes always challenged us. In the beginning of the trip, Freddy and Wally had erred by sending too many of us down into the fo'c'sle to eat at once, which left the deck empty in a couple of positions and for too long a time. Only two guys had been left on deck to keep the gurdy turning—hauling gear and catching fish. Fish to be cleaned piled up in the checker, and skates to be baited stacked up on deck. Coming back to work after a meal, it would take us an hour to catch up. From then on we learned to send only one man below to eat, assigning the meal break one at a time. The coordination of the crew began to click. I noticed that they were spending less time lighting cigarettes and grabbing cups of coffee.

One day while I worked on baiting a skate, the gear got hung up on the ocean bottom and Jack went forward to take over at the roller. I could see that Kaare had stopped coiling. Jack had stopped hauling with the gurdy and tried to free the hang-up by moving the *Grant* ahead through the water to pull the line tight. This delay gave me hope. I wanted so badly to bait a skate two-for-one. My slow speed at baiting skates continued to be my greatest struggle. Excited, I accelerated my pace. Turning to grab a hook, I glimpsed Chris, who stood behind me. He didn't say anything but just smiled, seeing what I was trying to do.

I finished the skate and tied the last knots just as the rollerman yelled out, "SKATE-OF-GEAR!" Even though the gurdy had been stopped for a long time, it still counted as two-for-one, a huge accomplishment for me. Exhausted, I pumped my hands into the air and yelled out so all could hear, "I did it! . . . I did it!" Chris grinned back and broke his silence. "Hey," he said. "It's our eleventh day today. It's hump day!" Whether we filled the boat or not, the trip was halfway done.

* * * *

The increasing weight of our catch was settling the *Grant* lower and lower in the water. When I helped the rollerman to pull in a big fish, I noticed that I didn't have to lean as far over the rail, nor did I have to lift as high. The boat was bringing the fish closer to me. At the beginning of the trip, the red paint that protected the bottom of the *Grant* could be seen as a foot-wide strip above the water's surface. Now the waterline was a foot under water. When the boat rolled, more water came through the scuppers and sloshed across the deck.

For Jack, a new problem arose. For him, fishing inside Sandman Reefs didn't outweigh the risk of going aground on a rock and losing the boat and, with it, our catch in the hold. He didn't want to push his luck. Jack told the crew, "We're makin' a move tonight . . . to the east, toward Kodiak." Closer to home.

We left Sandman Reefs with one week to go.

32

PRIMAL
EXTREMES

BEFORE DAYBREAK AT THE END OF OUR FIFTEENTH DAY OF
fishing, the crew had finished their work and left me up on the deck to
wash down the bait table. In a few minutes, I'd finished my work. Jack
saw that I was done and switched off the deck lights from the pilothouse,
transforming light to pitch black. Relieved to be done for the day, I went
blindly through the ritual of wrestling with my oilskins to remove them.
I watched Jack move as a shadow across the deck and disappear down
the fo'c'sle ladder, joining the rest of the crew for mug-up.

Darkness poured over me, blurring boundaries of boat and water,
ocean and sky. Mist hung in the air, hazing the stars above. My eyes
adjusted further, and I made out the aurora borealis hanging in a gauzy
curtain that spread across the northern horizon.

As I sat on the rail and struggled, trying to tug off my rain pants,
I heard a deep, sucking sound in the ocean behind me. I jerked my head
around to look at nothing but blackness. What I'd heard was strange.
When a small wave brushed against the side of a boat, it burbled and
went "ploop." With a ripple that collapsed, it made a "hiss." What was
this? It wasn't either of those sounds. I had heard a kind of a slurping

sound, like the slice of an oar pulled hard through the water. Then again, had I heard anything at all?

Out the corner of my eye, I detected the shape of what looked like a ghost moving in the water below me. Breathless, I wanted to run but couldn't. My pant legs were tangled tight around my ankles. Sitting on top of a skate, I had my oilskins pulled down. The hair of my arms and neck prickled and crawled.

"Get these fuckin' things off me," I hissed. "NOW, dammit!"

Just six feet behind me, the head of the ghost broke through the ocean's surface. Its mouth gaped open to roar like a lion, so loud and so close to me to make my clothing shake on my back. Sure that I was its prey, I screamed. Through some miracle, I got free of my pants and sprang across the deck, leaving my oilskins behind. I raced down the ladder into the fo'c'sle.

In my entry to the letter I wrote that night, I attempted to minimize my fright:

August 20—3,000 lb. This morning, I was taking off my oilskins after "hauling," when I hear this big "GrrowwLL" right by the boat. I about jumped out of my oilskins and then I found out it was a sea lion.

A couple of nights after my run in with the sea lion, we were making another move, cruising farther away from Sandman Reefs and closer to Kodiak. Chris had wakened me for my wheelwatch, my first watch in the night. I climbed out of the fo'c'sle to find blackness outside—no moon, no stars, and no aurora borealis. The red and green running-lights illuminated the cables that stretched up to the mast, and nothing else. The radiance of the tail mast light was lost into the night. The air was dead calm, with no wind.

Before entering the pilothouse, I was already concerned with my inability to see anything ahead of the boat. Joining Chris inside, I saw the dim light of the compass dial and looked for the green glow of the radar screen. It was dark, turned off.

"Why's it off?" I asked Chris, pointing at the radar.

"There is nothing to see," he said.

"So . . . why isn't it on? I can't see a damn thing."

"That's exactly why it's off," he said. "You don't need it. It's clear out.

There's nothing but water ahead of you. It's just open ocean. We're off-shore. Got it?"

"Seems kind of stupid, though, to leave it off."

"Well, think about it. If there was a boat out here, when it got within about eight miles of us, you'd see its mast light on the horizon. Which reminds me . . . no reading books at night. You've got to keep the lights off in here so you don't lose your night vision. You're on for two hours, then call Wally." He added, "Oh yeah, and pump out the fishhold at the end of your wheel."

Chris stepped out the door and slammed it shut.

I was on my own now and tried to adjust to my surroundings. I traveled blind and alone through the night. Outside, I saw the faint glow of the bow wave as it tumbled, illuminated green by the light on the starboard side. The big diesel groaned from the engine room below. In the pilothouse, an electric motor switched on and off, pumping the hydraulics for the rudder back and forth deep below the stern. I backed away from the windows to sit on the little seat, hoping like hell that we wouldn't hit anything.

During his watch, Chris had smoked in the pilothouse with the windows closed. I needed fresh air. I opened a window and stuck my head out—to breathe and to look around at nothing.

A light swell nudged the *Grant* and she rolled lazily. The mass of fish she carried had increased her inertia, softening the effect of the waves. Something about the way the boat moved was arousing.

A light breeze brushed across my face. It felt good, rustling the stubble of my whiskers, too thin and sparse to be considered a real beard. I had let it grow all summer with the plan of shaving once I got back home—but not until after I had fun toying with my mother, telling her that I was keeping it.

My excitement faded and I began to muse, entering into a state of reverie, and despite not being able to see beyond the bow of the boat, I found it beautiful outside. Green sparks of bioluminescent plankton streaked through the bow wave like shooting stars. Three porpoises began swimming around the bow and through the wake, beating their tails, making their turbulent path glow like a snake of cold light. It was then that the disturbance of the *Grant* passing through the water seemed to have started a chain reaction. Acres of ocean began to glow, for as far as I could see. It was magic. I couldn't believe my eyes.

I wanted to share this sight with someone. I struggled back and forth about whether to wake Chris or to leave him alone. I was sure he would be in a deep sleep by now. By reversing the situation with me in the bunk and not him, I overcame my hesitation and went down into the fo'c'sle to wake him.

I stirred Chris. He grimaced, looking confused.

"Ya gotta see this. The ocean's glowing!" I insisted.

He slipped on a couple of layers of clothes and followed me up to the deck.

"Wow! I've never seen anything like it!" he said. From the sound in his voice, I knew that any irritation he felt from being wakened was gone.

We gloried in the sight for a few minutes and then it faded away.

33

EVERYTHING
BUT SOUTH

THE STATE OF ALASKA IS MERCURIAL—NOT IN TERMS OF THE
god, but the planet. It feels like 1,300 degrees in the sun, and minus 300
in the shade. Alaska has the most of this and the least of that. It's got
the highest, the smallest, and the coldest. It's the biggest, the least pop-
ulated, and stretches the farthest north and west. Alaska extends over
three hundred miles into the Eastern Hemisphere, making it the east-
ernmost, too. Alaska is a colossal tract of emptiness, providing a basis
from which to appreciate its extremes, a dearth from which to gauge its
wealth. There is nothing half-assed about Alaska. It is the paradigm of
natural majesty.

In terms of marine organisms, above *and* below the surface, the Alaskan ocean is an aquatic savanna occupied by creatures of all shapes
and sizes, from plankton to whales. In the weeks that I spent in Alaska
this summer, I saw the clownlike grin of the orca, the lopsided head of
the sperm whale, and the explosive spout of several species of baleen
whales. A menagerie of marine mammals entertained me—fur seals,
harbor seals, and sea otters. Only the sea lions tormented my existence,
stealing halibut from our fishing lines.

Sea birds filled the skies, ranging in size from albatross with a seven-foot wingspan, to tiny petrels the size of sparrows. When I dressed fish and threw fish guts over the side, sea pigeons covered the water around the *Grant* in a solid blanket. They were brilliant flyers, seeming to communicate with the sea, taking commands from the waves. I marveled at watching them fly at breakneck speed mere inches above the waves—in fair weather and in storms.

Shearwaters, another sea bird, sometimes swarm in flocks estimated to number more than a million birds. When we fished around Sandman Reefs, I witnessed a flock of shearwaters so dense that it darkened the sky.

As a result of Jack's policy of using halibut guts as bait, I had the opportunity to study the stomach contents of our catch. During slow periods of fishing, I pretended that I performed autopsies on our halibut and found a diversity of sea creatures inside the guts, most too small to be capable of biting our large hooks. I found sculpins, gunnels, pollock, greenling, and even a baby octopus. The most unusual specimen I extricated from a halibut's stomach was a fleshy blob of a fish, the size of a fist, having a humanoid face and stubby fins. Jack identified it as a "smooth lumpsucker." In the gut of one fish, I found baby crabs—recently devoured. Some were still alive. I tossed them back into the sea, likely to be devoured a second time.

My experience of the second trip, same as the first, was scarred by weather systems that whorled across the North Pacific. In nice weather, I tolerated the work. In times of storms, I bent down and surrendered to the bullwhip of the wind. The ocean itself never scared me. I felt secure on the *Grant*. During the storms, however, I fought with real misery and grappled with a voice that screamed inside my head, "Why am I doing this to myself?"

After the storms, to emerge from the fo'c'sle and find the ocean clear of waves produced mood swings that felt narcotic. I had never felt a sense of relief so powerful. The euphoria healed my spirit and mended my body, helping me to press on.

The ocean forced me to feel. My senses ensured my safety—feeling, seeing, and hearing. It was imperative that I watch, feel, and listen to the ocean. I had to become more alert and responsive—and more alive—when on the ocean. My posture changed. I was more muscled and stood a little taller and more erect. The sinews in my back were responsible for most of this, but so was the knowledge that I had stepped up to this job and hadn't succumbed. I had discovered that I had a body of water.

The entire crew contributed to my success. Without them, I would have failed. Chris was my role model. He was a powerful young man. I knew that someday, with time and work, I could be strong like him and have as much endurance. Because he had recently graduated from being a greenhorn, the value of his lessons overshadowed the others. He remembered the obstacles to becoming a full-share fisherman and he showed me the tricks of the trade, helping me address my challenges and move on.

Kaare, the oldest man by twenty years, fatherly to the crew, grandfatherly to me, was a powerhouse of determination. He exhibited the mental tools of focus and concentration, skills in scant supply for a teenager. Quietly, he taught me a working philosophy, utilizing the wisdom of his even-keeled nature. He supported me in those times when I needed more clarity—in work and in life.

Freddy was the hustler on deck. Always the speedster, he showed me how to move fast, in moving my hands as well as in making quick decisions on deck. He made me question myself. Could I do more to produce, using energy and speed? And from Freddy I learned the value of a sense of humor in a bad situation.

For me to pick one crew member who embodied the role of father, it had to be Wally. His hard-ass style kept after me, hammering away with his discipline—though *my* father never did that. Wally saw my ability, prodded me, and forced me to find my own strengths.

Jack was different—he was the captain. When I went north to work on the *Grant*, I expected my uncle to captain the boat from the pilothouse. Chris had told me about some captains in the fleet who never left the shelter of the pilothouse in their work. They were known as "slipper skippers." I liked the way Jack captained the *Grant*. Our captain worked on deck, elbow to elbow with the crew and with me the greenhorn, through the good weather and the bad.

Though he worked on deck, Jack maintained a distance from me and the rest of the crew, separated by his role in the pilothouse. When I watched him, he gave me every indication that he'd been hardwired for this job of seaman, leader, and engineer. In the space of weeks, I had gained an inkling of the skills leading toward seamanship and leadership, but I hadn't witnessed the dirty work he performed in the engine room, and I still had no grasp of what it took for him to maintain and operate a vessel on the high seas. Given the man that I saw now, I supposed that at the start of his career, it must have been frustrating for him to be a crewman.

In the last week of the trip, I noticed that the crew had started a new habit of thanking Jack for finding good fishing and filling the boat. It seemed that after every time we hauled a string with a lot of fish, they'd say, "Nice job, Jack." I liked to hear this.

Before I worked for him, I would have described Jack in three words—sarcastic, funny, and needling. As my captain, I'd seen him be severe and kind, steadfast and yielding, giddy and serious, and compassionate and dispassionate. It fascinated me the way he could shift from being cautious, and in the next moment be cocksure to the absolute. I didn't know it at the time, but he was teaching me how to be a leader through his example.

Starting on the second trip at Sandman Reefs, my challenges in fishing expanded. The crew demanded that I work harder and pull my own weight as a fisherman. For the first time, I measured myself on an adult scale, and in doing so, I grew by leaps and bounds. I had learned so much in so short a time—about the ocean, fishing, and myself. In the good times, I loved my new job. In the bad times, I hated it. Still, I was hooked.

34

DECKLOAD

THE *GRANT* DRIFTED, IDLING BETWEEN TALL ISLANDS WITH sharp ridges and hollow craters, as if the hills had been gouged by a giant ice cream scoop. Same as last trip, Jack had set the gear in sheltered waters to make the work easier on our last day of fishing.

It seemed impossible that the hold was full of fish. I opened the small hatch to peer down, just to make sure. Yes. Instead of a black hole, I stared at a layer of ice just inches below. Where would we put today's catch?

The drudgery had ended last night when I baited my last hook. It was all downhill from now on. We would run to Kodiak, just shy of a two-day trip. We'd unload the boat and clean the hold, and then I'd shower and board the next plane to Anchorage to fly home. Though the last day of fishing was the busiest, I was eager to get to work, washing down the boat and sharpening hooks.

The crew celebrated while we worked. We had triumphed over the ocean, the waves and weather, the fish, and Sandman Reefs, and would earn a jackpot payout. But I witnessed more than that. For many seasons now, the halibut fleet's catch of fish had been poor. The crew knew that once we reached Kodiak, news of our success would spread over the

North Pacific Ocean, traveling via radio waves. A deckload trip would give an indication to the fleet that the halibut resource was showing signs of rebounding, something they all hoped for, fueling their spirit to stay in business rather than quit fishing. The effect of our catch would reach far beyond the deck of the *Grant*.

The crew was beside themselves, acting silly. I kept to myself, knowing that I had more to celebrate. Unlike them, I would be soon be traveling home. It was time for me to go back to school. I had served out my sentence on the *Grant*. After a layover, Jack and the crew would be going out for at least two more trips before their season finished. I'd be home asleep in bed while they hauled gear and baited hooks all night.

Home seemed so far away. I ached to feel the comfort of my bed affixed to the earth, unmoving. I wanted fresh food—milk and bread, fruits and vegetables. Strangely, Tony's beer and barstools had lost their attraction for me—same with the dancers. I wanted none of that. I wanted to hold Patti and press her body against mine.

The *Grant* inched through the sculpted islands, floating on glassy water. The gurdy groaned, "Rraau-rrraau-rrraau-rrraau. . . ." I kept busy, cutting and scraping, washing and scrubbing. At times I allowed my eyes to wander and look over the *Grant*'s railing to watch the landscape pass by.

Halfway through the day, at the end of the third string out of six, we'd cleaned enough halibut to fill up the checker on deck. The gurdy stopped, and with it, stopped the process of fishing. We had to do something with the fish.

"Okay guys. Get the hatch cover off," ordered Jack.

Lifting in unison, we raised the massive hatch cover from the combing and tipped up one side to rest on its edge. Then, and to my dismay, the crew kept pushing it over until the hatch cover lay nearly upside down against the rail, in complete violation of the superstition that they had drilled into me over the past seven weeks. Keep the hatch upright. I had adhered to their rules faithfully. This just seemed wrong. I shook my head.

Removing the hatch had opened up a hole in the deck four feet wide by six feet long. Seeing seawater slosh through the scuppers and wash right up to the base of the hatch combing, a barrier just thirty inches high, made me nervous. The hatch combing effectively dammed the water, yet I imagined thousands of gallons of water flowing into the hold, sinking the boat in seconds. It seemed like a precarious situation to me. I didn't have a clue what to do. For seven weeks, repetition had presided over my life—cycles of hauling and baiting gear, catching and cleaning fish, chopping and shoveling ice. This was new.

It seemed that Freddy and Wally didn't know what to do either, arguing over a course of action. Chris and Kaare hung back, and Jack didn't get involved in the discussion at all. Jack leaned out the pilothouse window, smiling. He took long drags on his cigarette, exhaling with clouds of smoke, enjoying the moment.

I saw that Freddy and Wally struggled with the situation, working together, trying to sort out alternatives for dealing with the novelty of the hold being full of fish—the best of problems for a fishing crew. To me, there seemed to be an air of incompetence here, topsy-turvy to the tight lines of organization that I normally experienced on deck.

At last, Freddy and Wally agreed on a plan. With tension in his voice, Wally shot out a series of orders, rapid-fire.

"Chris, get the scoop."

"Kaare, get the tarp."

"Lay the tarp out flat here . . . on top of the hatch cover."

"Dean, start shoveling and put the ice on top of the tarp . . . and don't let a fuckin' flake fall on the deck. We need it all."

While the others watched, I shoveled the ice out of the hold, down two feet to the level of the fish already iced in the slaughter-houses, depositing the ice on the overturned hatch cover. Of the 20,000 pounds of ice that had been blown into the hold in Kodiak, about 500 remained. We had used it all, packing it around our catch in the hold.

Freddy took my place down in the slaughterhouse and stood on top of the iced fish. Chris and Wally tried to kneel, slipping down the tilted hatch cover. They started a production line, stuffing the fish's pokes with ice and then flopping the fish down next to Freddy's feet. In no time, the stack of fish rose up so high that Freddy climbed out of the hold to stand on deck, leaning over the combing to arrange the fish in the hold.

In five minutes, I saw the sight that I had once thought impossible— the *Grant's* hold full to the brim with fish. Pride filled me, knowing the amount of work that had gone into that moment.

"That's it, guys," announced Freddy, "No more fish in the hold."

I looked into the checker and saw that a few fish remained on deck. Freddy, Chris, and Wally saw this too, turned to each other, pumped their fists into the sky, and yelled out, "Deeecckkk . . . loooaadd!!" Jack and Kaare just smiled.

In the last three strings, we finished our work of cleaning the boat and rotated through our jobs—coiling, filing hooks, and dressing fish. Done with baiting gear, we could take turns basking in the sun—a luxury. I lounged and reminisced, beginning to bid farewell to Alaska.

By the time the last bag and flag came aboard, our deckload had grown to reach 3,000 pounds. We stored these fish on deck in the checkers, kept moist and cool under several layers of canvas tarp. In all, we estimated our total catch at 73,000 pounds—a huge trip.

All work done, the day ended like a dream—in sunshine and on calm seas—a great feeling. My mind reeled with images from my summer's journey—flying to Kodiak and meeting my crewmates—the bars, the strippers, and getting drunk. I flashed on my exploration of the cannery at Alitak, then my first days at sea. I recalled my first storm and shuddered to think how miserable I'd been, only to be revived by the stop at Sand Point. I had to acknowledge that my success this summer may have hinged on something so small as the rupture of a hydraulic hose. I doubted that I could have survived a marathon of three straight weeks at sea without a break.

During the layover in Kodiak, I succumbed to the push and pull of guilty pleasure—tempted and seduced, only to become repulsed. An emptied wallet and a hangover helped me leave Kodiak, return to the ocean, and want to fish again. On the second trip, I finally became an advantage and not a burden to the crew. I was one of them. Starting at Sandman Reefs, I had become a true fisherman.

Jack put the boat in gear. The *Grant* ran for Kodiak at full speed. I looked ahead at going home now. Nothing could stop me.

* * * *

At 11:00 AM, a day and a half later, I rose from my bunk for a two-hour wheelwatch. This would be my last shift in the pilothouse before we arrived in Kodiak. Freddy bustled around the galley preparing lunch, clanging pots and pans without regard to those who tried to sleep.

I savored the warmth of the sun on my face as I strolled down the main deck to relieve Jack in the pilothouse. Judging by the hazy appearance of the shore passing on our port side, I guessed that we cruised about five miles offshore.

"Where are we?" I asked Jack, but before he had a chance to say, "In the middle of nowhere," I interrupted him with, "How long until we reach Kodiak?"

Seeing the eagerness in my eyes, Jack grinned. "We passed Shumagin Flats a long time ago," he said. "We're off Cape Barnabas, halfway up the southeastern shore of Kodiak Island."

His face became relaxed. "I'm going to bed. You've got the last watch. Call me in two hours. Our heading is forty-five degrees." Turning around, he ducked into his stateroom, stopping at the door to look back with a smile. "We're close to town. We'll be tied to the B&B's dock in about six hours," he said, closing the door behind him. Settling onto the little seat, I tried to relax, looking out an open window.

Bright sun cast the waves in blue. The *Grant* forged through six-foot swells topped with whitecaps that came at us bow on. The extra weight of the fish let the boat plow through the big lumps of water like they didn't exist, crushing them into white foam. The *Grant* had more momentum, seeming to transfer more power to me—the master of the helm. It felt like revenge on the waves. For me, it was a great day to quit the North Pacific.

Once we reached Kodiak, I'd be too busy unloading the boat to finish the last pages of my book. Before my two-hour duty ended, I wanted to finish *The Don Is Dead*, a Mafia novelette that piggybacked on the *The Godfather* craze. I started reading, trying to rush through the last chapters.

In minutes, my mind was lost in words. It felt good to relax and be distracted from the anxiety of going home. Reading made the time pass more quickly, too. Every so often, I glanced at the compass and monitored the scene around the boat. No land or boats ahead. Just blue waves and sky.

I had reached the climax of my book. Two Mafia thugs, Tony and Shorty, were about to set off a bomb . . .

Get set." Tony hissed.

Shorty put his hand on the switch. Tony waited while the second hand made one more sweep around his watch. "Now!"

Shorty closed the switch . . .

And the *Grant* smashed into the log.

PART SIX

Mayday

35

ORANGE

WE BOBBED ON THE OCEAN, CRAMMED INSIDE THE RAFT. With our backs pressed against the raft's outside tube, we sat in a circle with the slope of the canopy forcing our heads down. It was silent, except for the lapping of water outside. I stared down into the tangle of our legs. All was orange.

Sneaking a glance, I saw faces colored by shock and grief. We survived, but we had lost everything. Jack had abandoned his boat and his existence—the gear, charts, instruments, and volumes of logbooks. Together, we had abandoned a catch of fish worth a fortune, worth more than the sinking vessel that held it.

I didn't know it, but Jack had radioed a Mayday while I had been hanging over the side of the boat from my makeshift harness. He had established contact with the Coast Guard. He had also received an offer of help from a boat traveling down the coastline of Kodiak Island. The *Totem*, an 80-foot long crabber, was now heading to our position five miles offshore.

A sense of revulsion gripped me. I tried to keep from touching the others, but we were packed together. I thought of covering my face with my hands, but didn't, knowing somehow that when I removed them, nothing would have changed.

Looking out the opening of the raft, I squinted and saw a boat's mast sprout from the horizon. The hull of the boat, sitting at the level of the sea just like me, was still blocked from my view by the horizon. The mast grew taller until I could make out the shape of the *Totem*.

Coming at us at full speed, the crabber's big bow rose up to dwarf the raft. When I began to fear that the boat wouldn't stop, I saw a black cloud belch from her smokestack and heard the *Totem*'s engine roar. The boat slowed, turned broadside, and stopped in the water just twenty feet away. A crewman standing along the rail hailed out, "Heads up!" and threw a line. Jack leaned out the raft's opening, caught the line, and tied it to the raft. Together, the *Totem*'s crew pulled until the life raft bounced against the big boat.

Just outside the mouth of the raft, the side of the *Totem* whooshed up and down as she rolled, rising like a wooden curtain slowly at first, then flying up fast, making those of us close to the opening scramble to move away when the boat rolled back. The view from the raft changed colors as the *Totem* rolled, from red to black to white . . . then reversing, from white to black to red . . . and back . . . and so on. Algae hung from the *Totem*'s red bottom in green blobs.

When the *Totem* rolled toward us, the boat's pipe rail dropped down near the raft's mouth, stopped, and then accelerated ten feet straight up. I could see that in order for us to escape the raft, we would have to wait for the right moment, and grab at the pipe rail. But no one moved to get out. In the life raft, I knew that I was safe, but I wanted to get out . . . now. No one stopped me when I got to my knees, taking the initiative to go first.

I stuck my head out the canopy, blinking my eyes in the bright sun. I looked up the side of the *Totem* to the crew standing by, feeling anxiety build while I concentrated on calibrating the motion of the vessel. For this to work, I knew that I must remove any distraction from my mind. I quieted my fears and focused again on the movement of the pipe railing.

I let the rail swoop down several times within my reach, and then committed myself to grabbing the pipe, hanging on tight with both hands. The *Totem* snapped back and I shot out of the raft, slamming my legs and body against the side of the boat. Hands of the *Totem*'s crew latched onto my arms and waited until the boat reached the top of the roll, then pulled so hard that I flew airborne over the rail to land with both feet on the deck of the *Totem*. It felt like a rock compared to the flimsy life raft. The *Totem*'s crew plucked the others out, one by one, and then hauled the emptied raft out of the water to lie on deck. Seeing my crew safe, I was wordless with relief.

From the time we had jumped into the life raft, we had drifted only a half mile from the *Grant*, which barely floated nearby. Abandoned now, she was doomed. The bow deck, once eight feet above the waterline, had sunk to become awash. The stern had risen up to a high angle, making the boat appear ready to plunge to the seafloor at any moment.

In the surf that washed across the *Grant*'s main deck, I saw the brown hides of sea lions swimming, scavenging the halibut from our deckload. Seeing this made me feel sick. I lost any hope now that the ship might be saved. I watched and waited for the *Grant* to sink.

Jack walked to the stern of the *Totem*, leaned on the rail and stared out over the water to his boat. Jack shook with grief and sobbed as he bid the *Grant* farewell.

He was a shattered man.

36

THE THIRD
CHARM

LOOKING TO THE UPPER DECK, ONE OF THE *TOTEM*'S CREW hollered out, "Okay, Skip. Now what?" Standing behind the pilothouse, the captain yelled back, "The Coast Guard's sending a chopper. They're bringing pumps." It relieved me to hear that more help was on its way—and coming with a helicopter's speed.

From the moment that Skip, the *Totem*'s captain, had answered Jack's radio Mayday call, he had assumed the time-honored role of "sailors rendering assistance at sea." As the *Totem* approached the scene of the sinking *Grant*, Skip had maintained contact with the U.S. Coast Guard station on Kodiak Island, providing them details of our emergency so that the Coast Guard could strategize how best to help. The Coast Guard needed information of on-scene weather—sea condition, wind speed, sea swell height, and cloud cover—factors in determining whether a helicopter could be dispatched. Fortunately for us, the sky was clear and the wind moderate. With our crew now on board the *Totem*, attention could shift from saving lives to the possibility of saving the *Grant*.

Freddy and Wally volunteered to go back to the *Grant* upon hearing that the chopper carried pumps and that someone needed to be on board to operate them. Chris and Kaare gave no indication of volunteering, and Jack seemed in a state of shock, in no condition yet to return to his boat.

Skip put the *Totem* into gear to veer close to the *Grant*, and then, by jamming the engine hard into reverse gear, he made the *Totem's* stern whirl around to slam hard into the side of the *Grant*, making wooden planks of both vessels crunch with the collision. As the two boats came together, Freddy and Wally jumped, landing on the skates piled high on the *Grant's* stern. With the help of a wave, the *Totem* recoiled away, leaving Freddy and Wally alone on the sinking *Grant*. Far away on the horizon, I spied the Coast Guard helicopters, looking like a couple of hornets that homed in on us.

The first helicopter descended fast as it neared the *Grant*, with its turbine engine screaming and the rotor throwing up clouds of spray. Slowing in its descent, the chopper came so close to the *Grant* that I feared the twirling blades would strike her masts, swinging back and forth in the sky. The helicopter hovered, fixed in position over the *Grant's* stern.

The chopper's cargo door slid open and a coastguardsman leaned out, tossing one end of a line that swung down and struck the *Grant's* pilothouse. Wally swung his arm out and snatched the line before it whipped away in the tornado. At the other end of the line, a cylinder emerged from the chopper's side, hanging from a cable.

Freddy and Wally pulled on their end of the line as the cylinder was lowered from the helicopter. Compared to the big chopper in the sky above them, they looked like children with their arms reaching up, flailing.

The cylinder drew close. Freddy let go of the line and grabbed the cylinder, but just then the boat dropped into a trough between waves, forcing him to let go to keep from being lifted overboard. By the rise of the next swell, the cylinder had lowered enough so that he could hang on until it dropped to the deck. Freddy unclipped the cylinder and the chopper veered away, ascending fast to keep its cable from tangling in the *Grant's* masts and rigging.

Freddy and Wally set to work. They tore open the cylinder to find a pump inside and stretched a hose down into the engine room. I could see Wally pulling on a cord, like trying to start a lawnmower engine. He pulled and pulled.

Wally stopped for a moment, and even though I stood far away on the deck of the *Totem*, from the way his head jerked from side to side I could tell that he cursed at the pump. Though I couldn't hear him swear at it,

I shared his frustration. From the *Totem*, I cheered him on, "Keep going, Wally. Come on, dammit. START THE PUMP! ... GO!"

When Wally resumed pulling, I could see that he pulled slower now—he was losing strength. At this point, Freddy took over. He attacked the job, pulling like a wild man, but soon I noticed that Freddy, too, was reaching the limits of his endurance. My stomach tightened, and then I saw water gush from the pump's discharge hose. At last, the engine had started.

Clutching the rail of the *Totem*, I cheered out to the *Grant*, "That's the way, Freddy!"

Jack looked up—the sight of water coming out of the *Grant* seeming to shake him out of his daze.

To this point, Jack had let go of any hope of saving his boat. He knew that the damage in the bow had compromised all the compartments of the vessel, making the entire boat vulnerable to flooding. None of the five sections of the boat—the fo'c'sle, fishhold, engine room, aft cabin, and lazarette—functioned as watertight compartments. In his mind, it was a simply a matter of time before his boat disappeared beneath the waves.

Jack walked from the stern and sought out the skipper of the *Totem*, telling him that he wanted to go back on board the *Grant*. Soon after, I felt the rumble of the *Totem*'s engine being put into gear, and she swung around. This time, Skip steered the *Totem* to approach the *Grant* slowly. Then he stopped her, leaving twenty feet of water between the two boats.

Jack swung his leg over the rail, ready to jump. The *Totem* just drifted, rolling in the swell. Jack looked up to the pilothouse to see whether Skip intended to get closer, but Skip didn't respond. It appeared that Skip didn't want to risk his boat in another collision with the *Grant*.

Jack had no option—he had to swim for it. He spun around and jumped into the ocean. I watched him struggle to swim and guessed that he didn't feel the cold water, because he had other things on his mind—like saving his boat.

At seeing him jump, Wally and Freddy abandoned the pump to haul Jack out of the water. Three men had now returned to a sinking boat that lacked a life raft—the *Grant*'s sat on the *Totem*'s deck behind me.

The second helicopter made a drop to the *Grant* before leaving the scene—another pump, as well as a life raft packed in another cylinder. I guessed that the Coast Guard intended that those on the *Grant* have the life raft in case the *Grant* sank out from under them. The raft would give them something to jump into besides freezing water. But I could see that those on board the *Grant* had a different use in mind. Jack and Freddy fought their way to the fo'c'sle companionway, through the waves that crashed across the main deck, towing the raft canister behind them. They stuffed the raft's case down through the companionway, pushed it underwater into the flooded space, and pulled the cord to inflate it inside the fo'c'sle. The *Grant* needed more buoyancy in the bow of the boat. They had decided to use the raft to save the *Grant*—and not themselves.

The two helicopters departed for Kodiak, leaving the ocean quiet. The silence jarred with the rising tension I felt. More was at stake now. Jack, Wally, and Freddy were back on the *Grant*, their lives in danger. They had been rescued by the *Totem* once already. Could they be lucky enough to be rescued twice?

A small fishing boat—the *Lucky Lady*—appeared at the scene and pulled up to the *Grant*'s stern. They were traveling in the area and had heard Jack's Mayday. Their crew shouted over to the *Grant*, asking if they could help. Jack turned down their offer, telling them that the *Totem* had provided everything they needed. Nonetheless, the crew of the *Lucky Lady* threw a sack of supplies to them—a packet of sandwiches, candy bars, and a thermos bottle. Then Jack called out that they had cigarettes but no dry matches. Jack, Wally, and Freddy craved a smoke. The crew of the *Lucky Lady* tossed them a plastic bag of matches. Opening the thermos and taking a sip, Jack discovered that the coffee had been laced with brandy. The *Lucky Lady* departed and continued its trip down the coast of Kodiak Island, having done its bit to save the day.

Soon after the *Lucky Lady* left, Skip passed on word that a Coast Guard ship had been dispatched to assist in the rescue effort. The U.S. Coast Guard Cutter *Storis* would reach our position that evening, arriving in about nine hours. We all knew that the *Grant* could sink at any moment. We needed more help now, not nine hours from now.

The Coast Guard station in Kodiak radioed the *Totem*, ordering the skipper to take the *Grant* under tow and find shelter in Boulder Bay eight miles away on the southeastern coast of Kodiak Island. Skip obeyed the Coast Guard's order, but it was clear that he wasn't happy about it. Skip's "rendering of assistance" was turning into an ordeal.

The trip to Boulder Bay began with the *Totem* towing the *Grant* stern-first, using a pair of 200-foot shots of crab-pot buoyline that connected the two boats. The *Totem* used this buoyline—three-quarter-inch in diameter—when fishing for king crab, to pull up their crab pots from the seafloor.

A rig for towing normally involves a long section of heavy chain and cable, so heavy that both chain and cable sink down deep in the water in the shape of a bow, which provides a shock absorber under tension. The towline that we used was short and light and was stretched out in a straight line. Owing to the distance between the *Totem* and the *Grant*, the boats rode on different sea swells and never became synchronized in their rhythm on the waves. As the *Totem* lurched forward at the rise of a wave, I felt the pull of the *Grant* resisting with a powerful tug—a feeling that sickened me.

I was not the only one who felt uneasy. Not long after the *Totem* had started towing the *Grant*, Skip came out on deck and walked toward me, grasping a hunting knife. With a desperate look on his face, he stopped to address me, standing near to the lines that tied the *Grant* to his boat. With the blade raised up to the lines, he threatened, "I ought to just cut this and get it over with."

It made no sense that he singled me out for making his threat. It was more than clear that I was the junior member of the *Grant*'s crew. I held my tongue, certain that the tension of the rescue had driven Skip crazy. He started muttering in a rant that I couldn't understand, waving his arms and the knife in the air. Then he turned and stomped back to the pilothouse, perhaps having come to grips with the absurdity of cutting a line and abandoning three men on a sinking ship. Nonetheless, I could tell that Skip wanted this rescue over with, one way or another. I was in no position to argue with him. If the *Grant* sank while under tow, the weight of the *Grant* would pull the *Totem*'s stern under or roll her over. Either way it would sink the *Totem*. Also, if the towline snapped, the whiplash of the line could cut a person in two. Skip was acting like a madman, but he was looking out for his ship, his crew, and himself.

Around half past nine, the glow of the *Storis* appeared on the horizon, glittering as it approached, like a floating village on the sea. The cutter was massive, reminding me of the first time I saw a 747 parked outside Boeing's factory in Seattle. I estimated the *Storis* to be four times as long as the *Grant*.

The sheer size of the *Storis* raised my hopes. With a house fire, you want the fire department to show up. With a car wreck, the police and ambulance crew are your saviors. Finally, the Coast Guard was here to support us. I knew that they had come with people, skills, equipment, and lots of power. If the *Grant* could hang on and stay floating, it seemed that now we had it made. It was just a matter of time before the Coast Guard saved the *Grant*.

All the way from on board the *Totem*, I heard an intercom system on the *Storis*, the speakers blaring so loudly that I listened to the dialogue between the Coast Guard captain and Skip. The captain told Skip that he wanted a line heaved to the cutter to transfer the towline to the *Storis*, but, he added in another transmission, he didn't want to transfer any people while at sea. This left Chris, Kaare, and me to stay on board the *Totem*. Skip and his crew couldn't leave us and return to their fishing trip quite yet. After passing the *Grant's* towline to the *Storis*, the trio of vessels resumed the voyage to the shelter of Kodiak Island.

Around midnight, when we pulled into Boulder Bay, I lay sleeping on a bench in the *Totem's* galley and was awakened to board the *Storis*. Chris, Kaare, and I climbed a rope ladder up the side of the *Storis*. With the *Totem's* duty completed now, we bid farewell to its crew, expressing our thanks as they cast off their lines, seeing relief on their faces. I stood at the rail of the *Storis*, watching the *Totem* leave the harbor.

I hesitated, then I forced myself to walk to the other side of the cutter to go find the *Grant*, joining several crew of the *Storis* standing at the rail. I was shocked but not surprised at what I saw. Instead of the *Grant* being tied up with three lines—from the bow, the stern, and mid-deck—the *Grant* had over a dozen lines tying it to the *Storis*, looking like something out of Gulliver's Travels. The *Grant's* bow deck, once eight feet above the level of the sea, was awash. Three feet of water covered the main deck. Only her stern deck showed above the water. I looked to the main deck for any indication of our deckload of halibut, but not a single fish remained. The sea lions and the waves washing across the deck had taken them.

Looking haggard, Wally, Freddy, and Jack puffed on cigarettes, standing next to the *Grant's* pilothouse in the space where our stock of skates had once been stored.

Freddy shouted out, loud enough for all around to hear, "The third time's a charm, Jack."

"Huh?" Jack replied, not understanding him.

"Yeah. Third time's a charm. . . . The *Grant's* the third boat that's tried to sink from under me. The *Bonanza* sank, and so did the *Akutan*. The *Grant*'s not going to sink, Jack. . . . The third time's a charm!"

Seeing Jack force a smile made me smile, too, but it pained me to look at him.

Jack turned away to climb up a ladder to the main deck of the *Storis*. When Jack passed by me, it seemed like I was invisible to him.

Chris joined me and we stood at the rail of the *Storis*, working up theories to explain why the *Grant* continued to float. How the *Grant* had kept from sinking boggled my mind. The stern was floating, but the forward end of the boat didn't appear to have much remaining in the way of buoyancy. It looked like the bow could plunge underwater at any moment and never come up again.

I knew that the heaviest components of the *Grant* were the hull, the main engine, the anchor winch and anchor gear, and the 70,000 pounds of fish in the hold. Out of the water, 70,000 pounds of fish is a lot of weight. Compared to seawater, however, the weight of halibut is slight and nearly neutral. The construction of the *Grant* worked in her favor. Her planks and timbers were made of Douglas fir, a soft and low-density wood, with the low-density equating to buoyancy.

Ultimately, Chris and I guessed that the reason that the *Grant* floated up to now was due to several empty or nearly emptied tanks that lay hidden deep inside the vessel. By the end of our three-week fishing trip, we had depleted our reserves of fresh water and diesel fuel. Over 600 gallons of water and the better part of 2,000 gallons of fuel had been eliminated. Air or gas vapors were now in the tanks.

Flanking the engine room, the big steel tanks that were once filled with diesel were now nearly empty, which helped explain why the stern floated in spite of the *Grant*'s engine being the heaviest object in the boat. We couldn't ignore the value of the Coast Guard's life raft that we saw bulging out of the companionway, but Chris and I agreed that the fresh-water tank under the floor of the fo'c'sle was the key factor in answering why the *Grant* persisted. It was a big tank and it was filled with air. Yet, like the diesel tanks, we knew that this tank had a pipe for a vent, making the tank vulnerable to being flooded. We couldn't see the pipe above the water and feared that the tank was filling with water as we spoke.

We also had serious concerns about the integrity of the fo'c'sle water tank. From the way that our drinking water changed color when the seas got rough, we knew that rust pitted the walls of the tank. Compounding the question of the tank's integrity, this tank had not been designed for the purpose of keeping water out. It had been designed to *contain* water. Filled with air, this tank was under tremendous pressure now—under the weight of twelve feet of water above it—making the tank at grave risk of being crushed like a tin can. If the tank collapsed, the *Grant* would sink, snapping the lines to the *Storis* like threads.

37

THE
ILL-EQUIPPED
SHIP

NOT LONG AFTER MIDNIGHT, I STOOD ON THE *STORIS*, WHEN
Chief Warrant Officer Garry Perry, the coastguardsman in charge of the
deck, walked up and introduced himself to Jack. The officer expressed
his relief in the good fortune of our rescue. Jack replied with thanks, and
then asked, "You've got scuba gear, right?"

The officer's gaze dropped. "No, sir. I'm sorry, we don't."

Jack's jaw fell slack. He paused and shook his head, then asked, "A
diving mask?"

"No, we don't. Sorry, sir."

My uncle stepped away and started pacing. I knew a diver equipped
with a mask could locate the damage to the *Grant* within minutes, and
organize a strategy for the repair. I was stunned. Our saviors had arrived
at the scene, but they were not equipped for rescue.

Officer Perry, undaunted, spoke quickly and firmly to Jack. "Skipper,
we've got a wet suit."

Jack stopped pacing and glared back.

Officer Perry offered Jack a strategy for salvaging the *Grant*. To start,
locate the hole from the outside, plug the hole temporarily, raise the
Grant by pumping the fo'c'sle dry, and last, figure out a way to repair the

damage. To find the damage, Officer Perry offered a tactic somewhat akin to my idea of using a harness and dangling from the bow. Officer Perry summoned a crewman who stood nearby, carrying two wooden poles with a three-foot-square panel of canvas fastened between them at one end. Officer Perry called it a "fothering mat."

He explained, "We'll start by pumping water out of the fo'c'sle, then we'll locate the damaged area on the hull by moving the canvas of the fothering mat underwater back and forth along the planks." Gesturing with a sweeping motion of his arms, he continued, "One man handles each pole and they work together. When the canvas moves over the damaged area, the canvas will be sucked and held against the side of the boat by the water passing through the hole in your vessel, but the poles keep the canvas from being sucked entirely through. The canvas stops the flood of water, and then, we pump out the fo'c'sle and float the *Grant* again."

Jack replied, "Yeah, I didn't know what this thing was called, but I've heard of it before. Captain Cook used it, right?"

Officer Perry nodded.

"No way," I muttered to myself. Jack wasn't talking about Captain James Cook? He was 200 years dead. Certainly it couldn't be possible, but when I saw the fothering mat, saw its simplicity, I realized that they *were* speaking of Captain Cook of the British Royal Navy—circa the 1700s. Officer Perry had proposed that we use a device from the age of square-riggered sails—a primitive tool made of two sticks of wood and a swatch of canvas. I was dumbfounded.

Officer Perry said, "We've got two pumps here with six-inch-diameter suction hoses, and they're powerful."

"We've got to do something," barked Jack, impatience rising in his voice. "Let's get going."

Officer Perry left us to order the *Storis*'s crew to lower the suction hoses down to the *Grant*. One of the big hoses was extended into the *Grant*'s hull through the fo'c'sle companionway, squeezing it by the inflated raft, and the other down the fo'c'sle's ventilation funnel. I stood at the rail of the *Storis* with Chris and Kaare, looking down on the *Grant*.

Officer Perry yelled out the order. "Start the pumps!"

I had hoped, waited, and prayed for this moment.

The hoses lurched, became rigid, and within a few seconds, both hoses fell limp again.

"Stop the pumps!" hollered Officer Perry.

STOP? I kept myself from shouting out, "PUMP, GODDAMMIT!"

The *Storis*'s crew hefted the suction hoses, pulling them out of the *Grant* to discover that huge wads of debris, mostly fabric, had plugged the end of each hose, choking off the suction.

"Oh, shit!" Chris gasped. He leaned to me and whispered under his breath, "That's all of our stuff."

Chris and I exchanged glances, knowing what the Coast Guard did not. Our personal gear was down there—sleeping bags, pillows, sheets and blankets and three sea bags of clothes, plus our supply of food, all sloshing inside the fo'c'sle in a slurry, agitating like inside a giant washing machine. The coastguardsmen strained to pull the clumps of debris out of the hoses.

After they cleared hoses, the pumps started, only to be stopped, with the hoses plugged again. Over and over, the process started and stopped, each time with the hoses needing to be cleaned out, with my hopes rising and falling at each cycle. A surge of emotion pushed and pulled at me, making me feel near to breaking down from what I perceived to be an act of futility. I distracted myself by watching Jack and Wally work with the fothering mat, hanging onto the poles and groping blindly. They worked for an hour, but the hole eluded them. Nothing. Freddy took over.

Jack went over to Officer Perry. "I don't understand," Officer Perry said. "We should have found it by now." Jack just grumbled.

Wally and Freddy assisted the coastguardsmen in hauling up the suction hoses, clearing them of debris, and reinserting the hoses into the space of the fo'c'sle to resume pumping. Again and again, they repeated this process. Kaare, Chris, and I watched from the rail of the *Storis*. A few times, I recognized the shredded remains of my own clothing getting pulled out of the hoses.

Around 3:00 AM, Chris shouted over the noise of the pump, "Check it out. The bow has come up some." I looked and saw that the bow deck had surfaced to above sea level. The suction hoses had pumped for a longer period without clogging. For the first time since the log hit the *Grant*, the amount of water leaving the boat exceeded the amount that came in.

Seeing this thrilled me, but only for a moment. The pump clogged, the fo'c'sle flooded, and the *Grant* settled back into the ocean.

From that point on, the bow rose higher at each cycle before the pumping came to a halt and the boat sank down again. The pumps had cleared the fo'c'sle of most of the debris.

Around 4:00 AM, three hours into the pumping effort, the *Grant*'s bow raised up to be high in the water and the suction hoses began to slurp a mixture of air and water, making a sound like straws sucking up the last drops of the biggest milkshake ever.

I asked Jack if I could climb down to the *Grant*.

He didn't look toward me, but tipped his head and grunted.

Looking down inside the fo'c'sle, I saw the suction hoses resting on the fo'c'sle floor. A flood of seawater poured across the floorboards, the same image that had horrified me the day before. The only difference was that this time the big hoses were there to suck the water out, stabilizing the condition of the *Grant*. Still, the hole needed to be found and plugged from the outside. Wally and Freddy resumed working with the fothering mat, hoping to find the hole.

I lifted up the small hatch and looked into the hold, seeing no ice and nothing but fish. All the ice had been washed away, melted in the flooding. I knew that we needed to get the *Grant* to Kodiak as soon as possible to salvage our load of fish before it spoiled. Fat chance of that.

The idea of the fothering mat was so simple. Why wasn't it working? At least I assumed that the *Grant* had a hole in it. I wondered now if the collision with the log had dislodged entire planks from the hull.

I took a turn at the fothering mat, with Freddy hanging on to the other pole, and we experimented with different ways of hunting. I searched until my arms cramped up and I surrendered my pole to Wally.

A half hour later, as the sky lightened to the coming sun, Freddy yelled out, "I think we've got it! . . . Stop. Go down again, Wally."

They raised their poles up and then bent themselves down at the waist, extending the poles deep into the water. Their ability to move the poles became restricted at this point, indicating that a flood of water might be sucking the canvas tight against the hull, holding the canvas panel. In seconds, the *Storis*'s pumps started whining from pumping air—and no water. The fothering mat was stuck in place, stopping the flood of water.

"Stop the pumps," Officer Perry yelled out. Freddy and Wally slapped their hands together in a high five and then pounded each other on the back.

I quelled an urge to throw up.

38

"CHIPS"

JACK AND OFFICER PERRY BRAINSTORMED.

"We don't have scuba-gear, and the water is too cold anyway."

"You've got a wet suit on board, don't you?" snapped my uncle.

"Well, yeah. But we don't have any boots, gloves, or a hood."

Jack turned his head, eyes pinpoint. "Get the suit," he hissed.

In minutes, the Coast Guard's wet suit materialized on deck.

The crew of the *Storis* produced a long wooden plank from which they prepared a scaffold by tying lines to the board's ends and suspending it from the bow adjacent to the fothering mat, stuck to the *Grant's* hull.

An enlisted man from the cutter came up to Jack and saluted. He introduced himself as Damage Controlman First Class Hilliard "Chips" Middleton. Having once served a stint in the U.S. Army, Jack started to raise his arm up to return the salute but stopped himself.

Officer Perry said, "Officer Middleton here has volunteered to assist in the repairs and will be going into the water to secure the steel patch to your vessel." Though the wet suit would make a difference in protecting Chips from the cold, his hands and feet would be exposed to the icy water. His head would freeze, too, if he dove under water. Chips left with the wet suit and returned a few minutes later wearing it, with the black rubber stretched taut over his belly. With bare hands and feet, he shinnied down a rope ladder. Jack followed him to the scaffold.

Officer Perry joined the rest of us from the *Grant* to lean over the rail and look down at Chips and Jack on the scaffold.

Chips tested the water with his foot and winced. Without thinking, I scrunched up my face, too. I knew that Chips wasn't being dramatic. By getting drenched from heavy seas, I was all too familiar with the temperature of the North Pacific.

Chips sat down on the board and slipped into the black water, keeping his head above the surface. He pursed his lips, huffing and puffing, then piked over at the waist to dive down, disappearing into the ocean.

Within seconds, his head popped up at the surface. In a rush, he pulled himself out of the water to stand on the scaffold, with Jack helping him up. Wringing his hands, Chips panted and grunted, jumping from one foot to the other, aching from the piercing cold.

"I found it," he said, puffing his breath. "It's about a foot square. I groped down the poles and found it right away."

"Good . . . good," said Jack. "I'm glad it's not a whole plank."

"We'll get the engineers working on making a patch," said Officer Perry.

Jack climbed up the ladder, hustled toward the stern, and in seconds returned clutching a bundle of woolens—a hat, gloves, and socks—salvaged from his stateroom, the only dry place on the *Grant*. He tossed the articles down to Chips, who struggled to slip them on with shaking hands. A coastguardsman cried out from the deck of the *Storis*, holding a coffee cup. The coffee was passed down to Chips, who raised the cup, letting the steam pour over his face.

The engineers worked quickly in the bowels of the big cutter. In fifteen minutes, an engineer appeared on deck with sweat running down his face. He cradled an eighteen-inch square of quarter-inch steel plate in his hands. Holes had been drilled around the perimeter of the plate for fasteners—screws or nails—to attach the steel patch onto the wooden hull of the *Grant*.

Jack and Officer Perry came to an agreement that nails, rather than screws, would be used to secure the steel plate for the sixty-mile voyage to Kodiak. They decided that nails would be easier for Chips to hammer underwater—rather than having him try to manipulate screws and a screwdriver with hands paralyzed by the cold.

Chips descended back to the scaffold, looking tense. Using a line, Jack lowered the steel patch down to him. Chips found that he could just barely reach the hole in the *Grant* by lying face down on the scaffold and sticking his arms underwater. Chips began pounding nails into the hull with a small sledgehammer—a dreadful thing for me to watch. Chips's body went into contortions as he pounded, hindered by resistance from the icy ocean. To avoid exhaustion from working on the scaffold, he would jump into the water to work from a different angle, sometimes dunking his head underwater to gain leverage and pound from another position.

Watching him struggle made me angry. "If only we had a scuba tank," I fumed, but before I went further with blame, I beat myself with the thought, "If only I hadn't hit the log."

An hour later, Chips pounded the last nail into the hull of the *Grant*. He cut the canvas around the edges of the steel plate with a knife, freeing the wooden poles from the fothering mat. When he got off the platform, I could see that the skin of his hands and feet had turned blue with cold. The grimace on his face showed how he ached with pain.

Many hands pounded Chips on the back as he reached the deck of the *Grant*. Jack and Wally helped him up the ladder and back on board the *Storis*. A wave of gratitude with shouts and clapping engulfed him. He stopped briefly, delaying the hot shower that awaited him, to accept the praise due a true hero. A celebration erupted on deck. The saga of the rescue of the *Grant* had ended.

The sun climbed over the horizon.

39

CIGARETTES AND
TWICE-CAUGHT FISH

EIGHTEEN HOURS AGO, THE *GRANT* HAD HIT THE LOG. I HAD
been at the wheel and, in that instant, everything changed. Now, the
enormity of my mistake hit me hard. While the others celebrated, I hung
back and cursed myself. I questioned my decision to read my book.
I shamed myself for not having put it down . . . then I would have seen
that damned log. Heaped on top of shame, I felt guilt. The *Grant* struck a
floating log, and after the collision I had made up an excuse, telling Jack,
"We hit a deadhead." And Jack believed me. A deadhead is a log that
stands vertical and is mostly hidden below the water. I had lied to Jack,
trying to make myself look less at fault for not seeing the floating log.

Normally when someone screwed up, like a rollerman losing a fish
with a poorly placed gaff, Jack would lay into the crewman with a passion. Jack did nothing to me and said nothing. His inaction only made
me stew in my own disgrace.

It hurt to see my fellow crewmen look at me. Freddy gave me a sidelong glance. When my eyes caught his, Wally turned his head away.

Right now, I wanted Jack to lay into me, just for distraction, so that it
might help ease the pain of my humiliation.

I shuffled about the deck, gauging the cost of my mistake. The most obvious was the loss of the deckload—over 3,000 pounds of halibut. Not one fish remained. Those not washed overboard had been eaten by sea lions. The deckload had meant so much to all of us, much more than just money.

The *Grant* was saved, but not until we got to Kodiak would we find out how our load of fish had faired. Had the submersion in seawater harmed them . . . or worse yet, had diesel fuel or oil been washed from the engine room into the fishhold to contaminate the entire load?

My mind was locked into thoughts of ruin when the *Grant's* engine came back to life with a roar and revved up to high speed. A great cloud of white smoke billowed from the smokestack. How could that be? The engine had been submerged in seawater.

Jack came on deck from the engine room, his face glowing. "Looks like we'll be able to run to Kodiak under our own power, after all," he said, telling the others.

Turning to me, he asked, "Are you ready for a wheelwatch?"

I couldn't believe what he said. It seemed like a cruel joke. I'd failed in my job at the wheel of the *Grant* and had created this disaster. I knew that I'd fucked up from the time that I saw the log next to the boat. Was Jack toying with me?

Jack went on, "I'm relying on you. You're the most rested." I couldn't argue that point. The around-the-clock effort of saving the *Grant* had depleted Jack, Freddy, and Wally, leaving them in no shape for a wheelwatch. I had to admit to myself that I'd taken catnaps during the previous night. I had slept on the benches on board the *Totem* and *was* somewhat rested, certainly compared to them.

Kaare was out of commission. He had turned up drunk on deck, having consumed most of a bottle of vodka. I'd never seen a drop of liquor on the *Grant*. Where it had come from, I didn't know. On top of being drunk, Kaare was limping and his foot was wrapped up with rags. He had dropped the vodka bottle, breaking it, and had deeply cut his foot by stepping on the shards. That left Chris and me.

Same as me, Chris had napped during the rescue. I wanted to ask Jack, "What about Chris? Why not have him steer the boat?" But I didn't say anything. Jack wanted me to steer.

As Jack maneuvered the *Grant* away from the *Storis*, several of their crew looked down at us, waving goodbye. I waved back to them and then paused before joining Jack in the wheelhouse.

Everything in the pilothouse looked exactly the same as it did when we hit the log, except that my book was gone. Jack gave me the helm without hesitation, saying, "Wake me up abeam of Cape Chiniak. That's about two hours from town." He pointed his finger up the coastline to the northeast. "It'll take about four hours to get there. I'll take over for the last two hours of the run." He ducked into his stateroom, shut the door, and went to bed.

I steered the *Grant* by hand. The seawater that had flooded the engine room damaged many things beyond repair, including the autopilot. Grasping the wooden spokes once held by my grandfather, I steered the *Grant* through calm water, cruising alongside Kodiak Island in brilliant sunshine. The crew lay scattered about the deck, sleeping in the sun.

For the next four hours, I steered the *Grant* to swerve around each floating stick and piece of seaweed that I came upon, unsure of the integrity of the steel patch and fearful of hitting anything. I wished that I could boil up some coffee, but the flooding had destroyed the fo'c'sle stove. Grabbing a pack of cigarettes that Jack had left behind in the wheelhouse, I lit one. I hadn't smoked a cigarette all summer. I smoked it down to a butt and lit another.

Abeam of Cape Chiniak, I woke up Jack. He took over the helm and I withdrew to the stern. I lay down on the bait table and curled up in a ball, wrapping a blanket around me that the crew of the *Storis* had given us. I fell asleep.

* * * *

I woke up in Kodiak feeling relieved to find the *Grant* tied to the dock at B&B. I had slept while the others had docked the boat. The hatch cover was off and the unloading had already begun.

The *Grant* was safe and we were safe, and we still had the load of halibut on board, but the big question that remained was finding out whether B&B would buy our fish. Our cargo, potentially worth $70,000, had been submerged in seawater for hours. The ice that had preserved our fish had melted away soon after hitting the log. B&B's quality control inspectors hovered around the hatch combing of the *Grant*, looking into the hold for any indication of damage to the fish.

The first of our fish were hoisted up to B&B's dock in a net and the plant's workers descended on them for inspection, only to discover that they were in perfect condition. The same with the second net load. In fact, our fish looked better than those from a normal unloading because all the fish slime had been washed away. With each net load came a nod of a head, indicating good quality. My relief grew with each nod. Workers graded the fish by size and sent them straight into the freezers. With no ice in the hold to be shoveled out of the way, the unloading went fast and smoothly. Compared to the first trip, the fishhold was spotless.

40

REENTRY

AFTER ALL THE FISH WERE UNLOADED AND THE BOAT WAS cleaned and tidied up, Jack gathered the crew together. That is, except for Kaare, who was no longer in Kodiak. Jack had sent him away as soon as we had docked in Kodiak. Jack didn't want Kaare around the boat, drunk and disabled.

Jack told us what we already knew, that the damage to the *Grant* had put her out of commission for the rest of the fishing season. Jack proceeded to tell the crew that they were released from their jobs. Instead of being obliged to help Jack run the boat all the way back to Seattle, they could find jobs on other boats and finish *their* season. Jack told us that, with the help of his wife, he would be making the trip across the Gulf of Alaska and down the Inside Passage back to Seattle. This was good news and a relief for the crew—and an act of kindness by Jack. Still, all of the crew opted to fly home to Seattle to take a long-awaited layover from fishing, instead of staying in Kodiak.

It was afternoon by the time Freddy, Wally, Chris, and I caught a taxi to Kodiak airport. Because all my clothes had been lost in the flooding of the fo'c'sle, I wore clean clothes that Jack had given me. My stars-and-stripes ditty bag was my only possession to have survived from Seattle, having passed from the *Grant* to the *Totem*, to the *Storis*, and then back to the *Grant*. When the taxi accelerated toward the airport, I felt sick to my stomach.

While we waited in Kodiak's tiny air terminal, I took my one and only tongue-lashing for hitting the log. Chris and I had gone into the terminal's restroom and I made a casual comment to him about Freddy and Wally being grumpy after we'd salvaged the *Grant* and saved the load of fish.

Up until I had made this comment, the restroom had been silent. Then from inside a toilet stall Freddy's voice erupted, yelling, "Listen here, you little mother-fucker! I'll tell you why we're fuckin' grumpy. We don't have a fuckin' job!" He punched the stall's door open, rushed up to me, and continued, pointing his finger in my face. "You get it, you little prick! . . . Jeezus Christ! . . . I can't believe it!" He went on for a couple of minutes nonstop, summarizing the whole ordeal, punishing me with his jabbing finger and the blows of his words. I didn't react. Instead I became calmed, perhaps needing to hear his anger—and feel the anger of the crew. But that's all I ever got. Freddy, out of the whole crew, was the only one to chew me out. I had expected one of the others to lash out, like Jack or Wally, but least of all Freddy.

We flew together on the short flight to Anchorage. During our stop-over there, I walked alone through the airport terminal and came upon Kaare—a chance encounter. He was drunk, red-faced, and limping on his injured foot. The planes to Seattle had been full, so he was still waiting, trying to catch a flight. I was happy to see him. With his quick departure from Kodiak, I hadn't had a chance to say goodbye to him.

He smiled, greeting me, and then his face turned stern and sober. He grabbed me by the wrist, pulling me down close to look straight into his bloodshot eyes. Slurring his words, he said, "Hey, listen up, dammit. You're a good kid. You do well in school, now." He stopped and glared at me. "Do you hear me? . . . WELL, DO YA?!"

It startled me to be held by force. I met his gaze.

"Sure, Kaare. I'll do well," I assured him.

"Good. Now get the fuck outta here."

He threw my wrist down, turned, and hobbled away toward the bar. I never saw Kaare again.

As the jet raced down the Anchorage runway, gathering speed, I had the same feeling of nausea that I had experienced riding in the cab, only worse. The acceleration had made me feel sick again. It wasn't like me to get dizzy. I had become accustomed to the motion of the sea, being accelerated from one second to the next, up and down and from side to side, nonstop for weeks on end.

Something had changed. To make the feeling go away, I closed my eyes.

On the flight home, I prepared to adjust the way that I now behaved. Some changes would be less complicated than others. I spoke differently now and agonized over how I would control my new habit of swearing. In the past few weeks, cursing had become quite natural to me—and fun, too. I knew that Mom wouldn't tolerate this. Not even a slipup would pass by her without a reprimand.

It was time for my reentry into society. Time to go back to *normal*. Time to go home and rejoin my ordinary life with friends, family, and high school.

*   *   *   *

Mom and my brother met me at the airport that night. She watched her boy emerge from the jet-way, possibly the same jet-way through which I had departed seven weeks earlier. Her boy was different now. I wore strange clothing—articles that she had neither seen, washed, nor folded back at home. My face was gaunt and hollow. Exhaustion and work had made my eyes tired. My cheeks were ruddy from the weather. I sported a hint of a mustache and beard framed by a mane of wild hair. After we hugged, I noticed that she stood further from me now. My brother didn't say a word. I envied his bronze tan.

Mom said, "Patti called to say that she would come over after we got home."

Aside from the contents of my stars-and-stripes bag, all that I'd brought to Alaska had been lost. Bypassing the baggage area, we walked straight to Mom's car in the parking lot.

On the way home, Mom's driving seemed reckless to me. Her car seemed to have more power and speed. My fingers gripped the armrest as we zoomed down familiar roads. Closer to home, the reflection of the car's headlights flashed off rows of fence posts like a strobe light, hypnotizing me and nearly putting me to sleep. In a few minutes, the car slowed, turned onto our dirt driveway, and came to a halt in front of the Cabin. At the door, I dropped my stars-and-stripes bag and walked in.

To me, the Cabin seemed enormous. Compared to the fo'c'sle of the

Grant, it was big. But when I went into the bathroom, I became dizzy again and the walls seemed to collapse around me.

In just minutes after being home, I heard the sound of footsteps on the porch and then a knock at the door. I knew it had to be Patti. I hoped that she would be wearing the same halter top as when I last saw her. Opening the door and seeing her smile, it didn't matter to me that she wore a sweatshirt. Autumn had arrived early this year.

Embarrassed to hug Patti in front of my mother, I whisked out the door into the darkness to be with her. We held each other in the longest hug of our relationship. I didn't want to let go.

Patti and I kissed until breathless and then joined our friend Eric, who waited at the wheel of his parents' car sitting in my driveway. As Eric pulled the car out onto the street, both Eric and Patti started talking frantically, trying to tell me about the one thing on their minds—a new song, which my friends had nicknamed "Supersong."

"What is it?" I asked.

Patti tuned the radio to KJR at 950 kHz. I heard a piano and Elton John. She spun the dial to Seattle's other rock station, KOL—1300 kHz.

"That's it!" Eric and Patti cried out in unison.

With my arms wrapped around Patti, I heard for the first time a song that's now legendary, one of the finest inventions of rock music—Led Zeppelin's "Stairway to Heaven." I was never good at hearing lyrics and understood just a few lines of the song, but I loved the music. Chills went through my body as the song built in passion and momentum to the climax and ended quietly. My senses buzzed, even though I was exhausted.

The three of us drove down dark country roads and listened to more songs on the radio. Eric and Patti chattered, entertaining me by reciting the highlights of their summer. As much as I wanted this moment to last forever, I couldn't take any more . . . and I fell fast asleep.

* * * *

Two stories tall, Kent-Meridian High School had a large central court-yard. A hallway ran in a circuit between rows of classrooms, those facing the courtyard and those surrounding the building's perimeter, looking out into the world.

With the start of school, and like every day before the morning bell, chaos returned to the hallways as rivers of teenagers washed through them—one flowing clockwise and one counterclockwise, each mixing with the other.

On the first day of school and those that followed, the swirl of humanity scared me. Fearing to become sucked from an eddy like a little fish and thrown into the turmoil, I tried to remain concealed, still dazed and sleep deprived by the accident in Alaska. While I felt proud of my accomplishments, I also felt crippled by what I had done to the *Grant*— and to Jack and the crew.

Had I been able to articulate to my friends what I'd done that summer, I'm not sure that it would have mattered. I had convinced myself that they couldn't understand. My story was from the sea—and beyond their grasp.

41

GRAMPA'S
MODEL

NEARLY A MONTH HAD PASSED SINCE THE ACCIDENT, AND Jack had returned with his crippled boat to Seattle. The *Grant* loomed high above me now, cradled by the dry dock in the shipyard. Contrary to my mother's insistence that I go to school each and every day—whether I was sick, or not—on this school day, she had driven me into the city and dropped me off at Fishermen's Terminal. I had come today to help Jack.

I stood next to him and looked up to the round bottom of the *Grant's* hull. Given the circumstances, it seemed strange to me that Jack seemed very much at peace—not distressed for having lost the chance to finish the fishing season in Alaska. I supposed that he was grateful for having his boat.

Though I had seen my uncles' schooners out of the water many times, I gawked at them, seeing the bulk of the boat that was normally hidden underwater. Below the waterline, the schooner's hull appeared as if it were a living thing—the belly of a whale. Lines of furrows that marked rows of wooden planks ran back from the bow, wrapped around the stern, and passed underneath it—looking cetacean.

On the portside bow, just below the level of the waterline, I saw a blemish in the *Grant's* smooth shape. Small and square, it was the steel plate that Chips had nailed onto the hull. Looking closer, I saw the fringe of the fothering mat underneath. The exposed edge of the canvas had become tattered during the long voyage back to Seattle. This old invention had performed well as a gasket under steel.

High above us, a boat carpenter, a shipwright named George, stood high on a scaffold next to the hull. Using a crowbar, George worked to lever the steel patch free from the boat. A couple of tugs on the bar and the nails gave way and the plate fell, clanging as it hit the deck of the dry dock.

For the first time, I saw the damage caused by the log. A hole in the *Grant's* hull gaped six inches wide and ten inches long. George craned his neck to peer in through the hole, looking left and right.

"Jeezus, Jack," he called down. "It's just as I guessed. She took a perfect shot."

Jack peered up, rocking back and forth on his heels.

George said, "That fuckin' log hit her smack dab between the ribs, . . . and centered on just one plank, too. Fuck, oh dear. What's the chance of that?"

That log had beaten the odds to strike the *Grant* in the gap between her ribs. In my mind, a tempest began to rage.

"If that goddamn thing had hit just inches away—up or down, left or right—the worst you'd have to show for this would be some chipped paint?" George laughed.

I stood mute.

"Still can't believe it popped a hole in your boat," George added. "Must've been a big log."

It took George a half hour to extract each end of the damaged plank from the *Grant*, pounding, chipping, and levering it out of the way to where I could see inside the hull. The ribs of the *Grant* were revealed, columns of lumber, four by four inches thick and spaced only ten inches apart. Behind the ribs, I saw the inner layer of planking that had kept us from repairing the hole from inside the boat.

I fumed silently, stunned by the impact of George's remarks. Rows upon rows of perfectly good planks remained unscathed. Why would a boat sink when just one plank broke? I knew that such thinking made no sense, but still I felt furious. It didn't seem fair—that so much depended on so little.

George descended to the dock and walked off to the wood shop to prepare a new plank for the hull.

The purpose for me to skip school and be at the shipyard today was to help Jack restore the fo'c'sle to its original condition. Wally, Kaare, Chris, and Freddy were long gone. I didn't know how Kaare was doing, but I'd heard that the others had gotten jobs on other boats to complete what they needed to earn in the remaining season.

A few days ago, I'd received a check in the mail for the second trip. I was surprised to see the check—and that Jack and the crew had decided to pay me anything at all. It was made out for $750 and, along with the cash that Jack had handed over to me in Kodiak, I had made $1,000 for the two trips. In the same time, the rest of the crew had made over $10,000 each.

Instead of hiring a shipyard worker to clean up the mess from the accident, Jack had asked my mother to send me out to help for the day.

By the stink I smelled wafting from the boat, I could tell that the *Storis*'s pumps hadn't sucked all of the debris out of the fo'c'sle. More work remained.

Jack wrinkled his nose. He looked to me and said, "Sonny boy, I've got a helluva job for you. Follow me." As he turned away, I saw a smirk cross his face.

Jack ascended a tall ladder that leaned against the side of the *Grant* and I followed him up. Nearing the top, each step evoked a memory, starting with the sound and feeling of the impact. Reaching the top rung and seeing the companionway again made me remember the fear I'd seen in Freddy's face, and Jack sprinting across the deck in his underwear. My mind flashed to scenes of water—the fo'c'sle flooding and the torrent that washed across the deck. And the glow of orange while we sat in the raft.

Wavering, I followed Jack down into the fo'c'sle. The last time I had gone down these steps, the *Grant* had been sinking.

We stood, surveying the dark space. The stench made me want to gag. I couldn't believe that not too long ago, I had lived in this place.

"The flooding in the fo'c'sle caused quite a mess," said Jack, adding, "I'd like you to clean it up." He smiled, not trying to hide the satisfaction in his voice.

"Start in the forepeak," Jack ordered, pointing forward to where the galley benches came together in a V. "Here you go. Clean out these food lockers to start." He shoved a bucket and a flashlight at me and escaped up the ladder into fresh air.

I paused, then lifted the lid of the locker to look down. What I found disgusted me—churned-up food mixed with shredded clothing—all of it rotting. I put on some gloves and started digging with my hands down into the bilge through the muck. My old "dirty" job, cleaning the underside of the bait table, paled by comparison.

Before long, I was down on my knees, reaching into the locker, but I needed to go deeper. I got to my feet, bent over and stuck my torso headfirst into the locker, diving into the bilge, spreading my thighs out wide to catch myself and hang upside down, all the while reaching for more debris to put in my bucket. To empty it, I had to shimmy back out of the locker.

I descended again and again, groping blindly into the slop of moldy food mixed with bits of cloth, sometimes recognizing lumps of putrid flesh to be onions. I surprised myself for not throwing up, especially while I hung upside down.

About a half hour into my job, my hands hit something hard. Probing it with my fingers, I tugged several times at the object before it came free of the muck, making a sucking noise. It was a piece of wood. I climbed out of the locker to inspect it. It was about ten-inches long, six-inches wide, and three-inches thick. I brushed moldy onion from its surface to look at it more closely in the dim light.

Paint covered only one side; the other sides were bare. This was odd. On a boat, paint is typically used to protect as many surfaces as possible. I turned it over in my hands, inspecting it further. More curious to me, I found that the ends of the wood had been sheared off to be nearly clean and square, but not cut by a blade. Where the wood was broken off, slivers of wood lay over, all tilted in the same direction. I knew that only something with incredible force could do this to such a thick piece of wood and shear it off. I shuddered, suddenly becoming unaware of my surroundings. The smell, the garbage, and the dampness of the fo'c'sle fell away. I became focused on only one thing—this piece of wood.

The splinter stubs all lay in one direction, making it clear that the force shearing the plank had hit the side that was painted white. Could it be? I shuddered again. A close look at the painted side of the wood showed a large indentation and scratches in the paint. My skin began to crawl. No matter how hard I wanted to deny it, I had to face the truth— this chunk of wood had once been part of the hull of the *Grant*. I held in my hands the piece of planking that the floating log had shattered, letting the ocean rush in. Another tremor shook my body.

I had to find Jack.

I scrambled out of the fo'c'sle and down the ladder to the dock. Jack was walking back from the hardware store, carrying sacks of supplies. I reached out, showing him the piece of wood. He set down his bags to take it.

"What is it?" I asked him.

He looked it over, and in just a few seconds he said, "That's it, all right."

As though he rejected it, he shoved the piece of wood back to me. He snatched up his bags and walked away, disappearing into the machine shop.

I returned to the fo'c'sle and resumed my work, setting my discovery aside. That evening when my mom picked me up, I took the block of wood home with me—an odd memento of my summer in Alaska. At home, same as my grampa's model of the *Grant*, this artifact sat on a shelf to collect dust.

* * * *

One day, I came up with an idea and it struck me hard. This epiphany didn't follow the *Grant*'s accident by a few days. It took about five years, coming to me as I roamed the harbors of the West Coast, looking to buy my own fishing boat. During my second season, I had become a full-share fisherman. Fishing had been good to me. Twenty years old now, I had bought my first house and I lived near the docks in Ballard, where the *Grant* floated with the rest of the schooners in wintertime.

I couldn't let go of this idea. To make it happen, I had to drive to Mom's house to get my grandfather's model and the chunk of the *Grant*'s planking. Two years earlier, Mom had moved out of the Cabin and into a new house. I had helped her with the down payment.

Finding the shoebox, I looked inside and was taken aback to see that the model was in pieces. At once I set to work, assessing what needed to be done. When I returned home, I began the restoration of the model.

To begin, I dismantled what remained of the old model. I laid the pieces out on a table and began restoring each piece by filling scratches and holes with putty and working them over with sandpaper until the surfaces were smooth and ready to paint. Over time, my pace slowed, but not my desire.

I made trips to the hobby shop, scouring the store's shelves for paints that matched the colors of the *Grant*—the white of the hull and pilothouse; the black, tan, and gray of guards, masts, and deck equipment; and the blood-red of the bottom paint. I painted all of the model's pieces—the pilothouse, companionway, hatch combing, anchor winch, masts, boom, and the tiny gurdy and chute—dabbing the paint onto the model using the whiskers of a small brush. Before reassembling the pieces, I was impatient waiting for the paint to dry.

I spun out sewing thread from a spool to re-rig the masts and stays. The model took shape and began to look like the *Grant* again.

After working on the model for several weeks, I had completed the restoration, knotting and securing the final black strand of rigging. I picked up the model and held it, testing its weight and integrity.

The beauty of the model astonished me, yet I had merely restored the model's surfaces. What my grandfather had shaped and created from his memory was magnificent. Three generations of our family had now worked on and labored over this object, in miniature and in our real lives.

I made a trip to a trophy shop to have a small, brass label engraved. It read:

F/V GRANT
JACK G. KNUTSEN, CAPTAIN

Grampa had named his son after his boat. Jack's initial "G" stood for Grant.

I took the punched-out plank from the *Grant* and glued the brass label onto it, then drilled four holes in the top of the piece of wood. Into the holes, I inserted the shafts of four halibut fishhooks, inverted so the model could rest on the curve of the hooks. To finish, I glued tiny slivers of felt on the hooks where the model would rest on them, keeping the blood-red bottom paint from being scratched.

I picked up the model to place it on its new base, on that piece of planking that had nearly been the *Grant*'s undoing. Looking at this product of the efforts of my uncle and my grandfather, a floating log, and myself, I saw what symbolized to me the power and determination, the beauty and strength, and the vulnerability of life at sea.

And that is what I gave my uncle for Christmas.

GLOSSARY

NOTE: The vernacular of fishing can vary from vessel to vessel. Though most words and terms in this book apply to all vessels at sea, some of the language herein was specific to the *Grant* and her crew at this time. (All mariners refer to vessels in the feminine—i.e., "she" and "her.")

aft—*adv.* 1. at, close to, or toward the stern of a vessel. —*adj.* 2. situated toward, or at the stern.

aft deck—*n.* the aftmost deck. The *Grant's* aft deck was raised approximately two feet above the level of the main deck.

Agreement—*n.* see Set Line Agreement.

amidships—*adv.* 1. at, close to, or toward the middle part of a vessel. —*adj.* 2. situated toward, or at the middle part of a vessel.

amphipod—*n.* a very small, shrimplike crustacean.

anchor—*n.* 1. a device that digs into the ocean floor, thereby securing a vessel, fishing gear, or other objects. 2. fishing anchor, a 50 lb. anchor used to secure the ends of a string of gear to the ocean floor. —*v.* 3. the act of anchoring a vessel to the ocean floor with an anchor. 4. at anchor, the state of a vessel being anchored to the ocean floor.

astern—*adv.* 1. at, or toward the stern. —*v.* 2. a backward direction.

aurora borealis (or northern lights)—*n.* an atmospheric display of illuminated solar wind particles entering the magnetic field of the Earth's north pole.

autopilot—*n.* an electronic and mechanical device used to steer a vessel on a selected heading.

bag—*n*. a balloon-shaped, air-filled flotation device made of rubbery, heavy-duty plastic, usually red-orange in color.

bait tent—*n*. an area of the aft deck used for baiting gear, sheltered by a pipe-framed canvas tent structure.

becket—*n*. a short loop of nylon twine spliced into the ground line, used to attach the gangion (and hook) to the ground line.

bloodline—*n*. the kidney organ of a fish.

boom—*n*. a horizontal wooden pole positioned between the lower end of the *Grant*'s two masts.

bow—*n*. 1. the forwardmost part of a boat's hull. 2. the area of the deck at the forward end of a vessel.

bow deck—*n*. see forward deck.

breakwater—*n*. a rock or concrete structure or heavy wooden wall protecting a harbor from waves or sea swell.

buck—*v*. the up-and-down movement of a vessel in response to head-on collision with waves.

buoy—*n*. a buoyant device used for marking an area or object; also see bag.

buoyline—*n*. a line used for attaching buoy to anchor.

cabin—*n*. a room in a boat. A cabin can be either a space within a structure standing free on deck, or a space integrated into the hull of the vessel.

catch—*n*. 1. a quantity of fish harvested by a fishing boat, measured in either pounds or numbers of fish. —*v*. 2. to capture fish by means of fishing gear.

chart—*n*. a map of water area.

checker—*n*. adjoining spaces on the main deck, separated by low board walls, used for crew's work and for storage of catch.

checkerboard—*n*. boards, usually wooden, used to separate the main deck into checkers.

chopper—*n*. 1. a cleaver for cutting fish fillets into bait. 2. helicopter.

chute—*n*. a three-sided structure made of sheet metal, placed on the stern of the boat and used for setting gear into the water.

clean—*v*. to eviscerate a fish.

cleat—*n*. a device around which ropes are looped and secured.

cocomat—*n*. a large doormat made of coconut-husk fiber.

coiler—*n*. a crew person, in charge of handling and coiling ground line as it comes out of the gurdy.

coiler seat—*n*. a seat for the coiler, mounted to the side of the hatch combing.

comber—*n*. the part of a wave collapsing from being too tall and steep to support itself. Combers are the most dangerous waves at sea.

companionway—*n*. a vertically oriented passageway on a vessel, incorporated with a ladder. On the *Grant* and other schooners, the companionway also refers to the curved wooden structure above this access that prevents wind and sea spray from entering the fo'c'sle.

crossing—*n*. an area of lateral cross section in the hold, partitioned by boards; the *Grant's* hold had four crossings.

Deep Sea Fishermen's Union (or Union, DSFU)—*n.* the Seattle-based labor organization formed in 1912 by crewmen from longline vessels fishing for Pacific halibut. As of 2012, the DSFU remains active with issues concerning labor and longline fishery management of the North Pacific.

deliver—*v.* 1. to bring a quantity of fish to port. 2. to offload fish at a processing plant.

delivery—*n.* 1. the process of unloading fish from a vessel. 2. the quantity of fish unloaded from a fishing trip.

dory—*n.* a small wooden vessel, its hull shaped to allow multiple dories to be stacked, one within the other.

dress—*v.* to eviscerate a fish.

end—*n.* 1. the end of a string (or set). 2. the entire arrangement of bag, flag, buoy line and anchor rigged at the end of a string.

engine room—*n.* a below-deck area occupied by the main engine and other machinery.

fast–*adv.* firmly attached or tied.

fathom—*n.* a unit of measurement for depth, equal to six feet.

fathometer—*n.* an electronic instrument used to measure the depth from water's surface to seafloor.

fid—*n.* a hand tool for forcing apart strands of rope, so that two ropes may be woven, or spliced, together. A fid consists of a sharp metal spike with the head of the spike attached to a blunt wooden knob.

fishhold (or hold)—*n.* a below-deck area for holding fish.

Fishing Vessel Owners' Association (or Vessel Owners, FVOA)—*n.* the Seattle-based organization formed in 1914 by owners of halibut longline vessels fishing for Pacific halibut. As of 2012, the FVOA remains active with issues concerning labor and longline fishery management of the North Pacific.

flag (or flagpole)—*n.* a 12-foot bamboo pole with a red-orange flag on one end and steel (window) sash weights on the other, all of which is kept buoyant and floating upright by styrene floats attached to the middle of the pole.

fo'c'sle (fōk'səl)—*n.* the forecastle, a cabin within the bow of a boat, used on the *Grant* as crew's quarters and the galley.

fo'c'sle deck—*n.* see forward deck.

forepeak—*n.* a V-shaped area in the fo'c'sle farthest toward the bow.

forward—*adv.* 1. at, close to, or toward the bow of a vessel. 2. a direction forward.—*adj.* 3. situated toward, or at the bow.

forward deck (or fo'c'sle deck, or bow deck)—*n.* the forwardmost deck. The *Grant*'s forward deck was raised approximately two feet above the level of the main deck.

fother—*v.* to use a sheet of fabric, usually canvas, to cover a hole in the hull of a vessel, restricting the flow of incoming water.

fothering mat—*n.* fabric used in fothering a vessel; also see fother.

fluke—*n.* the part of an anchor that digs into the ocean floor.

gaff (or gaff hook)—*n.* a hooked shaft of steel with handle, used for controlling the movement of fish.

galley—*n.* the kitchen of a boat.

gangion—*n.* nylon twine–36 inches long–used to connect fishhook to groundline. Same as leader in sport fishing.

gangplank—*n.* a ramp leading to a dock or boat.

gear (or fishing gear)—*n.* the equipment used in longline fishing, equipment usually not attached to the vessel–skates, anchors, slipshots, and bags and flags.

greenhorn—*n.* a person new to fishing.

grub—*n.* food.

groundline—*n.* a rope used in longlining–usually 11/32 inch to 3/8 inch diameter–made of one of many synthetic fibers–nylon, dacron, polypropylene. In the 1960s fishermen in Alaska phased out the use of manila, a natural fiber used in longlining for centuries. The Grant's string (of fishing gear) consisted primarily of nylon line.

guard—*n.* a raised length of lumber made of ironbark wood, positioned above and parallel to the waterline of a wooden boat. The guard protects the softer wood of the hull from abrasion and wear against docks and log pilings.

guardrail—*n.* a raised wooden board or bumper used for protecting a vessel's hull from impact or abrasion, usually made of dense hardwood like ironbark.

gurdy—*n.* a mechanical device used to haul or pull in lines.

hatch—*n.* 1. an opening allowing for vertical passage from one area to another. 2. a removable cover for the opening that allows vertical passage from one area to another. 3. the smaller and topmost of the two hatch-combing covers.

hatch combing—*n.* a rectangular opening in the main deck, with raised sides above the level of the deck to keep water from entering the area below.

hatch cover—*n.* the hatch covering the hatch combing to the fishhold.

haul out—*v.* to remove a vessel from the water, typically via a rail-borne cradle or a floating dry dock.

header—*n.* a crew person tasked with decapitating fish.

heavy seas—*n.* a sea condition of large waves.

highliner—*n.* a fishing captain with a reputation for catching large quantities of fish.

hold (or fishhold)—*n.* a below-deck area for holding fish.

horns—*n.* see roller.

hull—*n.* the watertight external structure of a boat.

ice—*n.* 1. shave ice used in the preservation of fresh fish products. —*v.* 2. to pack the gut cavity of fish with ice and arrange fish in the fish-hold.

Iron Mike—*n.* slang for a vessel's autopilot.

jog—*n.* abbreviation for jog turn, jog watch. —*v.* the act of steering a boat to maintain position, usually performed while in heavy seas.

jog turn (or jog watch)—*n.* watch duty for maintaining a vessel's position.

lazarette—*n.* the most aft space within the hull of a boat that usually houses the vessel's steering gear and top of the rudder post. The lazarette on the *Grant* was also used as a cabin for crew's quarters.

line—*n.* rope used in longlining.

list—*n.* the lateral angle of a boat floating at rest.

main deck (also well deck)—*n.* the central of the three decks on the *Grant*.

masts—*n.* heavy poles of wood raised upright and supported by steel cable rigging. The *Grant*'s two masts supported a boom and held an array of navigation lights and radio antennaes.

mike—*n.* 1. a radio microphone. 2. short for "Iron Mike"; also see Iron Mike.

mug-up—*n.* a light meal or snack, usually eaten before bedtime.

oilskins—*n.* fishermen's raingear. The name comes from olden days when raingear was made from canvas material saturated with oil.

pallet—*n.* a wooden foundation on which cargo is stacked.

peak—*n.* see forepeak.

pen—*n.* one of several partitioned areas of the fishhold.

penboard—*n.* a wooden board used to partition the fishhold into pens.

piling (or pile)—*n.* a log driven into the earth, usually for making docks or breakwaters, or for marking shallow objects or channels.

pilothouse (or wheelhouse)—*n.* the above-deck structure housing a vessel's steering and engine controls and navigation equipment. On the *Grant*, the captain's stateroom was in the after part of the pilothouse.

pipe rail (or pipe railing)—*n.* a short railing made of heavy steel pipe. On the *Grant*, many of the wooden railings had a pipe railing on top.

poke—*n.* the gut cavity of a fish.

poop deck—*n.* see aft deck.

port—*n.* 1. on or along the left-hand side of a vessel (facing forward). 2. town with harbor capable of servicing vessels. —*adj.* 3. a direction to the left (facing forward).

quarter—*n.* one of the four areas around a boat, between its major axes–fore and aft, port and starboard.

quarters—*n.* the living or sleeping area for crew on a boat.

queer one (or queer wave)—*n.* a wave significantly larger than the norm in a given sea condition. The frequency of queer waves in storms is low but regular. See also rogue wave.

raft—*v.* 1. to tie a boat to the outer side of another boat, or several boats, tied to a dock. —*n.* 2. a floating object.

rail (or railing)—*n.* the uppermost, above-deck structure of the side of the boat. Also called the gunwale in other nautical disciplines. In addition to the wooden railing that surrounded the perimeter of the *Grant*'s decks, pipe railings were in some places attached to the top of the wooden rail. These were convenient for tying to or for hanging on to by hand.

rigging—*n.* the cables, clamps, and pulleys used to support and work with the masts and boom.

rogue wave—*n.* an extremely large and rare wave found in storms; also see queer one.

roller—*n.* a rolling steel cylinder over which the groundline passes when coming aboard, over the starboard rail. Horns extending up from below each side of the cylinder limit the movement of the groundline, preventing it from slipping off the sides of the roller.

run—*v.* 1. to travel between places.—*n.* 2. the journey between places.

running—*v.* traveling at sea.

sand flea—n. see amphipod.

scraper—*n.* a hand tool used for scraping the bloodline (kidney organ) from the gut cavity of a fish.

scoop—*n.* a gallon-size metal scoop with a handle, used for icing fish.

scuppers—*n.* openings through the bottom of the solid railing of a vessel, allowing the escape of water from the deck.

sea legs—*n.* 1. familiarity with aspects of life at sea. 2. the ability to adjust one's balance to the motion of a rolling boat.

seawall—*n.* see breakwater.

seine—*n.* a net specially designed to encircle schooling fish.

seiner (or seine boat)—*n.* a vessel employing a seine net to capture fish.

set—*n.* 1. see string (def. 1).—*v.* 2. the act and process of putting longline gear in the water.

Set Line Agreement (or Agreement)—*n.* the contract first established in 1916 between the Deep Sea Fishermen's Union (Seattle) and the Fishing Vessel Owners' Association (Seattle) for the purpose of establishing a share- and incentive-based partnership between longline crews and vessel owners. As of 2012, the Agreement is still in effect with little change over the years.

shack—*n.* bait for fishing made from fish caught on longline.

shitter—*n.* the space on a boat housing a toilet.

shot—*n.* a standardized length of line. On the *Grant,* a "shot" of buoy-line equaled 50 fathoms.

side pen—*n.* an area in the fishhold partitioned by boards, positioned along outer sides of the fishhold.

skate—*n.* a groundline used in longline fishing—1,800-foot-long section of 3-strand nylon with 70 hooks attached by 36-inch-long gangions attached to short loops (beckets) spliced into the groundline, spaced apart in 26-foot increments. Hook spacing has varied greatly, by boat and over time. Also see "skate of gear"

skatebottom—*n.* a square of canvas with short ropes attached to tie around a skate of gear.

"skate of gear"—*n.* a call used by the rollerman, directed to the other crew to indicate that the end of a skate was coming aboard.

skiff—*n.* a small boat capable of being rowed.

slaughterhouse—*n.* the central pens in the fishhold, partitioned by boards; see side pen.

slipshot—*n.* a length of line used for connecting fishing anchor to fishing gear.

snarl—*n.* 1. a large tangle of fishing gear. —*v.* 2. the act of tangling fishing gear.

soaker—*n.* a halibut larger than average, usually over 100 pounds.

sou'wester—*n.* a waterproof hat used in fishing, shaped with a surrounding bill extended in the back.

splice—*v.* 1. to join sections of rope together by interweaving strands of one rope through the other. —*n.* 2. the woven or knotted joint of two sections of line.

splicer—*n.* see fid.

starboard—*n.* 1. on or along the right-hand side of a vessel (facing the forward end of the boat).—*adj.* 2. a direction to the right (facing the forward end of the boat).

stateroom—*n.* a bedroom on a boat.

stem—*n.* 1. the forwardmost, vertically oriented part of the bow. 2. (in a wooden boat) the piece of heavy lumber at the stem.

stern—*n.* 1. the most rearward part of a vessel. 2. the area of the deck at the rearward end of a vessel.

stern deck—*n.* see aft deck.

string—*n.* 1. a number of skates (~ 8–12) tied together in series, with each end anchored, and with buoys and flags marking each end's position on the ocean's surface. 2. the entire stock of fishing gear on board a longline vessel.

swell—*n.* the underlying and longest sea wave. Swells can originate from locations many hundreds of miles away.

tender—*n.* a vessel used to service other boats. Salmon tenders assist boats fishing for salmon by buying their fish and by selling them supplies like food and fuel.

trawl—*n.* a long, conical net towed through water to capture fish. —*v.* to tow such a net through the water.

trawler—*n.* a vessel employing a trawl to catch fish.

trip—*n.* the complete cycle and period of time of preparing to fish, traveling to the fishing grounds, and returning to port to deliver the fish caught.

troll—*v.* to tow a line with a hook attached at its terminal end, luring fish to bite the hook by movement of the lure or the smell of bait.

troller—*n.* a vessel employing troll gear to catch fish.

twine—*n.* small diameter line (< 1/8th inch), usually braided.

Union—*n.* see Deep Sea Fishermen's Union.

uptown—*n.* the commercial area of a fishing town housing drinking establishments.

Vessel Owners—*n.* see Fishing Vessel Owners' Association.

watch—*n.* a designated amount of time that a crewman is "at the wheel" (in control of the boat); see wheelwatch.

waterline—*n.* 1. the line where a vessel's bottom paint and hull paint meet. 2. the line on a hull at the level of water in which the boat floats.

well deck—*n.* see main deck.

wheel—*n.* 1. A vessel's steering wheel. 2. A vessel's propeller. 3. wheel-watch.

wheelhouse—*n.* see pilothouse.

wheelwatch—*n.* the duty for steering a vessel, also called watch.

wind-speed categories (1 knot [kt] = ~1.1 mph) —
small craft warning	– >25 knots
gale-force winds	– >35 knots
storm-force winds	– >55 knots
hurricane-force winds	– >75 knots

wind wave—*n.* the wave caused by the disturbance of wind at that location. Wind waves occur on top of ocean swells. The direction that wind waves travel can be independent of the direction of the underlying swell.

ACKNOWLEDGMENTS

I begin by addressing the boys who read this book—those like me with a "D" in eighth-grade English. In my eighth-grade year, my teacher crushed my spirit—I learned that I could not write. Don't give up, boys. Write journals, on paper. You can burn them later if you like.

During the summer of 1972, I wrote journal letters—three to family and two to my girlfriend. Nothing was burned; all survived. By the time I finished high school, writing still didn't matter much to me—fishing was paying well—and I lacked the passion and desire to complete a paragraph. In 1978, everything changed—I bought my own boat. As a twenty-two-year-old captain, my new job required the command of words, and the focus, for composing letters to the federal government. I learned to express my concerns with fisheries policies and management following my uncle's example. Jack—a master of sarcasm—had a wicked pen.

Returning to the University of Washington in the 1990s, I defeated my demons of composition under April Denonno's tutelage. At the School of Aquatic and Fishery Science, I studied and wrote for two extraordinary professors among many—Thomas Quinn, virtuoso of scientific writing, and Ray Hilborn, standard-bearer of the scientific method.

For a project like *Four Thousand Hooks*, I needed additional guidance. At the start, Gloria Kempton provided gentle encouragement. I appreciated Christine Ummel-Hosler's instruction on a primitive version of the manuscript. Jennifer McCord, my writing coach through several rewrites, made a tremendous impact, asking "WHY?" and demanding "MORE." The most difficult directive of all? "ELIMINATE THE WORD 'WAS' FROM THE MANUSCRIPT." Ouch.

The following people reviewed details in the book: Robert M. Thorstenson Sr., Alaska seafood industry icon, examined the chapter on the salmon cannery line. Woody Knebel verified my recollections of the Alitak cannery. Scott Smiley, chief scientist and resident of Kodiak, checked for accuracy concerning the Kodiak community circa 1972 and confirmed the lowdown as well; before his career as a seafood scientist, he had worked as a bouncer in Kodiak's infamous bar, "The Beachcomber."

My greatest fan, my wife, Lori, witnessed the entire process of this book. She edited, cajoled, soothed, and, when invited, argued with me. When my morale flagged as I scoured the Web for potential agents, she took over and addressed scores of query letters. She never gave up—thank you, Lori.

While I wrote, my daughter, Courtney, frightened me regularly. She intimidated me with her talent in writing and editing—the evidence is scattered throughout these pages. For this, I am grateful. I thank my son, Connor, for insisting that more emphasis be placed on our family's Norwegian heritage. For that, I am proud—Ja, vi eisker dette landet.

I thank the ever-mindful Richard Fields for sharing his wisdom. James Thayer, the most published author in my neighborhood, advised me to "WRITE WITH FORCE. GIVE POWER TO YOUR VOICE." Yoda, he is—though much taller. Through several readings—one in a pinch—Cheryl Gordon proved the worth of her insight. Editor Lael Morgan's visceral reaction to sample chapters inspired me to retrench; for several months, I slashed away at the fat, cutting deep. The evolution of *Four Thousand Hooks* was finally complete.

Jerry and Pat Henderson pointed me to UW Press, where my new, lean manuscript found a friend in Executive Editor Lorri Hagman. Passing through the door of the Press, I was infected by the enthusiasm of editors Marilyn Trueblood and Gretchen Van Meter and received beautiful gifts from Ashley Saleeba and Thomas Eykemans, who generously invited me into the design process. I write these acknowledgments before the book's release, and I have only begun to value the marketing work of Alice Herbig, Rachael Levay, and Phoebe Daniels, expert handlers of my boyish energy.

I thank artist and biologist Joan Forsberg for the diagrams of the *Grant*, and Pat Grant, teacher, historian, and artist, for the Alaska maps. I also thank two anonymous reviewers for their critique of the preliminary manuscript.

For allowing me to share the words that were meant for her alone, I am very grateful to Patti. I bow to my mother and her oceanic courage in letting me go north at fifteen. Thank you, Mom, for keeping the letters.

In conclusion, I am indebted to my Uncle Peter for many things, one of which is a list of suggested readings that he supplied over the five years I spent on his boat, the Northern. The classics he recommended came at a precious time in my life and affected the way I think in a profound way.

And you boys, empower yourselves, Grab a piece of paper. Write something on it. And keep on writing. You just never know, you might have your words published.

Dean Adams
Seattle